UNCOMPLICATED CALCULUS

UNCOMPLICATED CALCULUS

The Beginner's Guide to Mastering the Mathematical Language of Change

JAY BEZIE

Tide House

CONTENTS

A Brief History of Calculus

Spanning centuries, cultures, and continents, the history and development of calculus is a fascinating but often overlooked story. Attracting many great minds in history, it is a tale of human curiosity, persistence, and the inherent need to understand the world around us.

The origins of calculus can be traced back to ancient times, when the basic principles of calculus were used in areas such as astronomy, engineering, and land surveying. The ancient Egyptians and Greeks, for instance, used proto-calculus concepts to solve problems involving areas and volumes to build temples and town infrastructure.

Leap forward to the 17th Century and two of history's most renowned mathematicians, Sir Isaac Newton of Britain, and Gottfried Wilhelm Leibniz of Germany, independently invented the modern field of calculus. This era is known as the *Age of Enlightenment*, a historical time of rapid progress in science and mathematics.

The renowned physicist, mathematician, and astronomer, Isaac Newton, developed his version of calculus to explain the motion of celestial bodies (which refers to any natural object or entity that exists in space, including Earth and the solar system), formulating laws of motion and universal gravitation. His work in calculus is often referred to as "fluxions", which is an older term referring to the methods and concepts he employed and that laid the foundation for the branch of mathematics we now commonly refer to as calculus.

At the same time in Germany, Gottfried Wilhelm Leibniz, a philosopher, logician, and mathematician, developed a more systematic approach, introducing integral and differential calculus as separate entities, as well as the notation used prominently today in calculus, such as 'd' for differentials and the integral symbol '∫'.

This dual but independent discovery led to a significant dispute, known as the "priority dispute", over which man and nation invented calculus first. Today, though, both men are credited as co-founders of modern calculus, reflecting their contributions and the different perspectives and notations they brought to this field.

During the 18th and 19th centuries, the field of calculus was developed further and formalized by influential mathematicians such as Augustin-Louis Cauchy, Karl

Weierstrass, and Bernhard Riemann. They worked on the foundations of calculus, including the formal definitions of limits, continuity, differentiability, and integrability, which are all essential aspects of modern calculus and are covered in detail later on in this book.

Calculus remains a powerful mathematical tool today. It not only underpins modern physics but also plays a crucial role in other disciplines such as engineering, economics, computer science, and even biology.

In upcoming chapters, we will journey through the world of calculus, exploring its principles and common applications, and on the way demystifying this important and far-reaching branch of mathematics. Whether you are just starting in calculus or taking it for years, this book will gently walk through the fundamentals without losing you in complex equations along the way.

As we begin, remember that every new concept and term will be defined and explained clearly to make your journey through calculus as smooth and easy as possible.

Why Learn Calculus?

Used to solve complex and important problems in the real world, calculus is a powerful branch of mathematics that focuses on the study of motion, including continuity, limits, definite integrals, and derivatives. While these terms may sound foreign and intimidating, they are essentially different techniques for studying the motion of physical objects like vehicles and cannonballs or more abstract concepts such as economic growth.

The field of calculus can be elementary enough for high school textbooks but also ample for advanced application in many professional fields including data science, physics, engineering, aviation, and economics. Calculus is often perceived as a challenging subject, and if this is true, why is it so vital that we push through the perceived difficulty to learn it? To answer that question, let's first define what calculus implies. At its core, calculus is the mathematics of change and motion. It provides a framework for understanding how things change and how quantities are accumulated.

Second, the importance of calculus extends far beyond the classroom and underlies many of the fields and technologies that modern society relies upon today, including the following categories.

1. **Physics and engineering:** As a language of physics, calculus is crucial for understanding the laws of motion, electromagnetism, and thermodynamics. Whether it's calculating stress on bridges or optimizing an electrical circuit, calculus also plays a vital role in engineering.
2. **Economics and finance:** In economics, calculus is used to model and predict economic change, optimize profit and loss, and calculate rates such as interest and growth. In finance, it's used for quantifying risk and return, portfolio optimization, and in pricing derivative securities (stocks, bonds, commodities, currencies, interest rates, or market indexes).
3. **Medicine and biology:** Calculus aids in the understanding of biological processes, modeling population dynamics, determining rates of infection spread, and analyzing medical imaging, among others.
4. **Computer Science and information technology:** In computer science, calculus is essential for machine learning, data analysis, graphics, and algorithm

3

analysis. It helps in creating more efficient algorithms and understanding their computational complexity.

5. **Environmental studies:** Calculus helps us model and predict population growth, the spread of diseases, and changes in ecosystems. It also assists in interpreting data on pollutant levels and climate change.

6. **Space exploration:** From launching satellites to planning interplanetary missions, calculus aids in optimizing flight paths, understanding gravitational fields, and designing spacecraft.

Outside of professional contexts, the mindset you develop when learning calculus —including problem-solving, logical reasoning, and the ability to handle abstract concepts—also holds broad applications. These skills are useful in many life situations and decision-making scenarios, promoting critical thinking and analytical abilities.

Learning calculus will seem daunting at first. But, with the right perspective, it will become clear why calculus is not only a fundamental academic subject but also a powerful tool that opens a wealth of opportunities across various fields. As you progress through this book, I aim to not just teach you calculus but also illustrate its importance, beauty, and practical applications.

Calculus in Real Life

While the previous section underlined the wide range of fields where calculus is used and applied, it may still seem a little abstract at first look. To bring it closer to your everyday experiences, let's look at some practical examples of calculus in action.

1. **Medicine and predicting the spread of disease:** During an outbreak, public health officials can use calculus-based models to predict how quickly a disease will spread. These models account for numerous factors such as rate of infection, recovery rate, and population density. This calculus-informed modeling informs public health strategies and helps manage the outbreak more effectively.

2. **Business and maximizing profits:** Companies often use calculus to maximize their profits. Suppose a company wants to produce a certain product. The cost of production and the revenue generated from selling the product are both functions of the number of units produced and sold. Calculus then helps determine the optimal number of units to produce and sell to maximize profit, a principle referred to as "profit maximization".

3. **Engineering and optimizing designs:** In civil engineering, calculus is used to calculate the optimal materials needed for building structures. For example, to construct a dome or a bridge, engineers use calculus to determine the curve that will use the least amount of material and yet remain structurally sound. This principle is known as "minimizing surface area".

4. **Navigation and GPS systems:** The GPS system in your smartphone or car relies on calculus for accurate operation. Calculus helps in predicting and adjusting the route based on speed, direction, and time. It also helps in accurately measuring the distance between different geographical locations, a process known as "trilateration".

5. **Computer graphics:** The smooth movement of characters in video games and animations is made possible by calculus. When animators want a character to move from one position to another, they use calculus to create smooth transitions and realistic movements.

6. **Weather forecasting:** Calculus helps meteorologists predict the weather by modeling the atmosphere, ocean currents, and temperature changes. By analyzing these factors, they can forecast weather patterns, track storms, and provide early warnings for severe weather conditions.

These six applications are just a few real-world examples. The beauty of calculus lies in its universality and applicability. By understanding calculus, you gain a powerful tool to comprehend and influence the world around you. As you delve further into the subject, always remember that calculus is not just about abstract numbers and symbols but also about real-world applications and problem-solving.

Prerequisites

Learning calculus is a rewarding journey, but like all journeys, it's important to be prepared. Let's talk first about the foundational knowledge that will make your learning experience smoother and more productive.

Understanding basic algebra is the first step. You should be comfortable with manipulating algebraic expressions and equations. These include operations such as addition, subtraction, multiplication, and division, as well as powers and roots. You should also be familiar with factoring, simplifying expressions, and solving linear and quadratic equations.

Beyond algebra, a solid grasp of functions is crucial. A function, in mathematical terms, is a relation between a set of inputs and a set of permissible outputs, where each input is related to exactly one output. This encompasses understanding the concept of function domains and ranges, as well as being able to interpret and sketch the graph of a function.

A simple function could be a relationship between the numbers you input into a vending machine, and the snack that comes out.

As an example, suppose you're standing in front of a vending machine with a keypad for numerical inputs. This machine has only three slots for snacks, and each slot has a unique number associated with it: 1, 2, or 3.

Slot 1 contains a **bag of chips**.

Slot 2 contains a **chocolate bar**.

Slot 3 contains a **pack of gum**.

This vending machine can be considered a function. Here's why:

Inputs and outputs: The numbers you press on the keypad (1, 2, or 3) are the inputs, and the snack that comes out is the output.

Each input maps to one output: According to the definition, in a function, each input is related to exactly one output. This is true for our vending machine: if you press '1', you always get a bag of chips. If you press '2', you always get a chocolate bar. If you press '3', you receive a pack of gum. Regardless of how many times you press '1', you won't get anything other than a bag of chips. Therefore, each input is mapped and linked to exactly one potential output.

Function domain and range: The domain of a function is the set of all possible input values, while the range is the set of all possible output values. In our vending machine example, the domain is {1, 2, 3}, which is the numbers you can press, and the range is {bag of chips, chocolate bar, pack of gum}, which are the snacks you can receive.

Interpret and sketch the graph: If we were to draw this relationship, we could sketch a simple graph. On one axis, we have our inputs (1, 2, 3), and on the other, we have our outputs (bag of chips, chocolate bar, pack of gum). We would then create a unique bar from each input to its corresponding output.

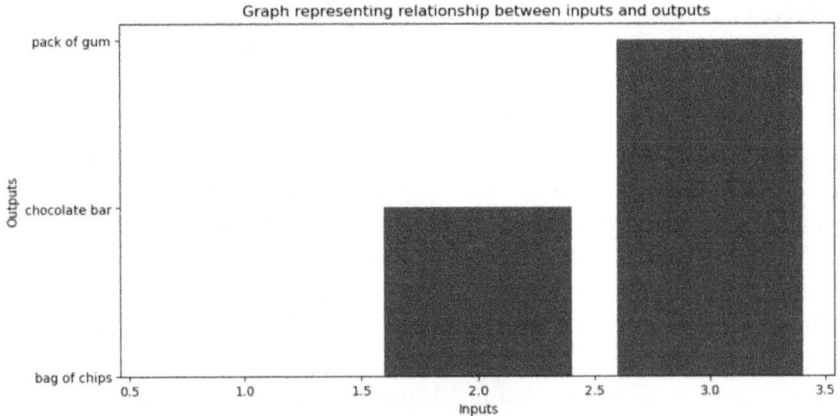

Figure 1: Graph of vending machine inputs (horizontal) and outputs (vertical)

To recap, a function is a relationship where each input corresponds to exactly one output. The vending machine example illustrates this definition: by selecting a number, you will receive a particular snack associated with that number. The concept of function domains and ranges was also demonstrated: the domain being the set of numbers you could choose (inputs), and the range being the set of snacks you could receive (outputs). Finally, we touched on interpreting this relationship visually by sketching a graph, allowing us to see how each input is tied to its corresponding output. As you continue exploring calculus, you'll notice the recurring theme of functions and the essential role they play.

Geometry, too, is a significant player in the world of calculus. You should be familiar with the basics of Euclidean geometry, including the properties of lines, angles, circles, and common geometric figures such as triangles and rectangles. More specifically, you'll need to understand how to calculate areas and volumes as these concepts lay the groundwork for integral calculus.

Trigonometry, the study of relationships between angles and side lengths of triangles, is another important component. Familiarity with sine, cosine, and tangent functions, as well as the unit circle, will play a key role, particularly when we delve into differential calculus and the study of periodic functions.

Specifically, sine is the ratio of the opposite side to the hypotenuse, cosine is the ratio of the adjacent side to the hypotenuse (the longest side of a right-angled triangle), and tangent is the ratio of the opposite side to the adjacent side. The unit circle is a circle with a radius of 1 centered at the origin of a coordinate plane, and it's used to define these trigonometric functions for all real numbers, not just for angles between 0 and 90 degrees, as each point on the unit circle corresponds to an angle and has coordinates $(\cos\theta, \sin\theta)$, making it a vital tool in trigonometry.

A basic understanding of mathematical logic, such as the ability to comprehend mathematical statements and their negations, can also be beneficial. This will assist in grasping definitions and theorems that we will encounter.

Finally, you will need a healthy dose of curiosity and persistence. As with learning any complex subject, you may find some concepts in calculus challenging. But don't be discouraged. The key to mastering calculus, as with any subject, is consistent practice and patient dedication.

Don't worry if you're not completely comfortable with all these prerequisites. This book will review these topics as they arise in the context of learning calculus. But the more familiar you are with these topics, the easier your journey through calculus will be.

How to Use This Book

This book has been designed to make your journey into calculus as seamless and enjoyable as possible. Here are a few strategies to get the most out of this book.

Firstly, embrace an active learning approach. This means not just passively reading the content, but engaging with it. As you move through the book, take the time to solve the exercises provided at the end of each chapter. These are designed to reinforce the concepts and techniques introduced in the chapter, and practicing them is key to internalizing and mastering these topics.

It's also important to pace yourself. Calculus is a complex subject, and trying to learn it too quickly can lead to confusion and frustration. Plan to spend regular, scheduled study sessions with this book, rather than trying to cram everything at once.

A slower, steady pace will give you the time to digest new concepts and practice new skills, and you will also find that you can retain the information better.

Don't be afraid to revisit earlier material. Calculus is a cumulative subject and each new topic builds on previous ones. If you're struggling with a new concept, it may be because you need to reinforce an earlier one. Feel free to flip back and review as needed. Each chapter is designed to be a self-contained unit that can be studied on its own, so you can easily refresh your memory or clarify any confusion.

While studying, if you encounter a new term that's not clear, you can refer to the glossary at the end of the book. It will give you concise definitions of the key terms used in calculus.

Finally, remember to engage your curiosity and enjoy the process of learning. Calculus is a fascinating subject that has shaped our understanding of the world in profound ways. Along the way, take the time to appreciate the elegance of the concepts and the power of the techniques you're learning and observe how calculus might be used in designs and applications around you.

CHAPTER 1

Fundamentals

1.1. Introduction to Mathematical Functions

Functions are the backbone of calculus, and understanding them is your first step. To define a function, let's start by envisioning a machine. Imagine you have a machine with an input slot and an output slot. You feed something into the input, the machine performs its operation, and out comes the result. In mathematics, we often call this machine a function.

A function takes an input, often called an 'argument' or 'variable', and maps it to an output, known as the 'value' of the function. This relationship is typically expressed as $f(x) = y$, where f is the function, x is the input, and y is the output. This notation reads as "f of x equals y".

One important characteristic of a function is that it produces the same output for the same input. Consider a function as a recipe. If you follow the same recipe (the function) with the same ingredients (the input), you will always produce the same dish (the output).

To demonstrate, imagine a simple function outlining the act of making a cup of coffee. The function takes an input (quantity of coffee grounds, amount of water, and amount of sugar) and returns an output (a cup of coffee).

If we consider one set of inputs, such as 2 tablespoons of coffee grounds, 200 ml of water, and 1 teaspoon of sugar, we insert these inputs into our function, which is the brewing process. This process might involve steps like heating the water, adding the coffee and sugar, and stirring. The result is a cup of coffee that tastes a certain way based on those inputs.

The key point here is consistency: if we provide the same inputs to our coffee-making function, we will get the same cup of coffee every time. This holds even if we're making coffee in different locations, or at different times. The output is solely determined by the inputs we provide, and this consistency is a key feature of mathematical functions.

We might express the same process in mathematical terms, using the variable x to represent the input and f(x) to represent the function. For our coffee-making function, f(2 tablespoons of coffee, 200 ml water, 1 teaspoon sugar) = 1 cup of coffee. Regardless of when or where we compute this function, as long as we use the same inputs, we'll always get the same cup of coffee as our output.

This principle extends to more complex functions in mathematics, such as those used in algebra or calculus. The variables and functions may be abstract, but the principle remains the same: given the same inputs, a function will always produce the same outputs.

Next, let's talk about the 'domain' and the 'range'. The domain of a function is the complete set of possible inputs, i.e., the values that you can feed into your machine. On the other hand, the range of a function is the set of possible outputs, the results that your machine can produce.

For example, let's consider the function $f(x) = x^2$ (reads as "f of x equals x squared"), where x is any real number. The domain of this function is all real numbers[1], since you can square any real number. The range, however, is only positive real numbers and zero, since squaring a real number never yields a negative result.

At this point, it's worth noting that not every relationship between variables is a function. For a relationship to qualify as a function, every input must be associated with exactly one output. This is known as the 'vertical line test' in graphical terms. If you can draw a vertical line that intersects the graph at more than one point, the graph does not represent a function, as it means that a single input (x-coordinate) corresponds to multiple outputs (y-coordinates).

To demonstrate this, imagine you have two sets of data. In the first set, you record the amount of hours you study each day and in the second set, you record the score you get on a test at the end of the week.

If we assume that the amount of hours you study directly affects the score you get on the test, then for every input (hours studied), there is exactly one output (test score). This means if you study for 3 hours, you get a specific score, let's say 85. If you repeat this process and study for 3 hours on different days, you consistently get a score of 85. This is a function because one input (3 hours of studying) gives you exactly one output (a score of 85).

However, if you have a different set of data where the time of day you start studying affects your mood, it could be that at 3 PM, you sometimes feel happy, sometimes feel neutral, and sometimes feel tired. If we graph this relationship with the time of day on the x-axis and mood on the y-axis, a vertical line at 3 PM would intersect the graph at three points, corresponding to the three different moods. This is not a function because a single input (3 PM) corresponds to multiple outputs (different moods).

So, in the first case, the vertical line test is passed (each vertical line intersects the graph at exactly one point), making the relationship a function. In the second case, the vertical line test failed (a vertical line intersects the graph at more than one point), making the relationship not a function.

Next, functions come in various types and each type has its own set of rules. For instance, there are linear functions, polynomial functions, trigonometric functions, exponential functions, and logarithmic functions, each with unique characteristics and applications, as summarized below.

Linear functions: These are functions of the form of $f(x) = mx + b$, where m is the slope and b is the y-intercept. In a linear function, the rate of change is constant. This means that for any equal increment in x, there is a consistent increment in $f(x)$. The graph of a linear function is a straight line, hence the name. These types of functions are seen in phenomena with constant rates of change, such as speed.

Polynomial functions: These are functions that can be expressed in the form $f(x) = a_n x^n + a_{n-1} x^{n-1} + ... + a_2 x^2 + a_1 x + a_0$. The highest power, n, is called the degree of the polynomial. Polynomial functions include linear functions (degree 1), quadratic functions (degree 2), cubic functions (degree 3), and so on. The behavior of a polynomial function largely depends on its degree and coefficients.

Trigonometric functions: As mentioned earlier, these include the sine, cosine, and tangent functions, which originate from the study of triangles and the unit circle. They are periodic (i.e., they repeat their values in regular intervals or periods). Trigonometric functions are crucial in modeling and understanding cycles, rotations, waves, and vibrations.

Exponential functions: Functions of the form $f(x) = a*b^x$, where the variable x is the exponent. The number b is the base and it must be greater than 0. Exponential functions model phenomena with constant relative rates of change, such as compound interest or population growth.

Logarithmic functions: The inverse of exponential functions, with the form $f(x) = \log_b(x)$, where b is the base. Logarithmic functions are useful in many areas of science and engineering, including when dealing with quantities that vary over a large range, like in the case of earthquakes (Richter scale) or sound intensity (decibels).

Each of these types of functions has its own characteristics and understanding functions is key to understanding calculus, as calculus involves the manipulation of functions. For instance, in relation to differential calculus, we will explore how functions change, while in integral calculus, we'll study how to accumulate quantities.

As we delve further into the intricacies of mathematical functions, remember that the function is your machine, transforming inputs into outputs. It's your tool to navigate the vast landscape of calculus. In the forthcoming sections, we'll dive deeper into different types of functions, their properties, and how to manipulate them—a necessary skill for mastering calculus.

1.2. Basic Algebra Refresher

Before we proceed further into calculus, it's important to review some fundamental algebraic concepts, as algebra is the language in which calculus is expressed.

One of the core components of algebra is the notion of a variable. A variable is a symbol that represents an unknown quantity. The power of variables is that they don't stand for a specific known number, but rather for "any number". This allows us to generalize, which is extremely important in mathematics. Instead of having to create and solve a different equation for every specific case, we can write down an equation involving variables that can handle infinitely many cases all at once. Variables therefore allow us to write general expressions and formulas that can be applied to different specific cases.

In practice, variables are often denoted using letters of the alphabet. For example, the letters x, y, and z are frequently used to denote variables. When you see an equation like $y = 3x + 2$, both y and x are variables. In this context, x is often referred to as the independent variable, while y is the dependent variable. This is because the value of y depends on whatever value we choose for x.

In the equation $y = 3x + 2$, we can substitute any real number for x, and the equation will give us the corresponding value of y. This equation, in fact, describes a rule for getting from x (input) to y (output), and this rule is the same no matter what specific number x might be.

Another example could be a formula for the area of a rectangle, which is $A = l * w$, where A is the area, l is the length, and w is the width. These variables allow us to calculate the area for any rectangle just by plugging in the specific values of length and width for that rectangle.

In calculus, variables take on even more depth because we start dealing with rates of change, which involves small changes in variables, and limits, which involve quantities that get arbitrarily close to some value.

These concepts would be difficult, if not impossible, to handle without the concept of a variable.

Next, let's briefly explore algebraic operations and how they are used with variables. The most common operations are addition, subtraction, multiplication, and division. With variables, these operations can be applied in a generalized manner. For instance, if we have the variable x, we can write expressions like x + 2 or 3x − 7, where the exact values depend on the value of x.

A crucial concept tied to operations is the idea of an 'equation'. An equation is a statement of equality between two expressions. For instance, x + 2 = 3 is an equation. The goal with equations is often to find the values of the variables that make the equation true, a process known as 'solving' the equation.

Now, let's delve into the concept of a 'function' from an algebraic perspective. We can think of functions as special kinds of equations. While an equation might have more than one value for a given variable, a function has exactly one output for each input.

Consider the equation y = x^2 + 2x + 1. Here, for every value of x, there's one value of y. If you replace x with 1, y becomes 4. Replace x with 2, and y becomes 9. Hence, the equation also describes a function, often written as f(x) = x^2 + 2x + 1.

However, not all equations can be functions. As mentioned earlier, an equation that can assign more than one value of y for a single value of x does not define a function. For instance, the equation x = y^2 has more than one value of y for a given x (except for x=0), and hence does not define a function.

Example

x = 2^2, = 4

x = -2^2, = 4

(The value of x is 4 for both y = -2 and y = 2.)

Understanding algebraic expressions and equations, and being able to manipulate them, is an important skill in calculus. This is because much of calculus involves working with functions, which, as we've seen, are algebraic equations with a specific restriction, which refers to a condition or limitation imposed on a variable, function, or equation. It sets boundaries or specific requirements that must be satisfied for the variable or equation to be valid.

1.3. Coordinate Systems and Graphs

A pivotal step in understanding calculus is visualizing mathematical functions, and the primary tool for this is the Cartesian coordinate system, coupled with the art of graphing.

The Cartesian coordinate system, named after the French mathematician René Descartes, provides a framework that allows us to represent numerical relationships visually. It comprises two number lines, called axes, which intersect perpendicularly at a point called the origin. The horizontal line is termed the x-axis, while the vertical one is the y-axis.

Every point on this grid can be identified by a unique pair of numerical values, denoted as (x, y), termed as 'coordinates'. The x-coordinate, or 'abscissa', tells us how far to move left or right from the origin, while the y-coordinate, or 'ordinate', tells us how far to move up or down.

In this context, we introduce the concept of a 'graph'. The graph of a function is the set of all points (x, y) in the Cartesian plane such that y is equal to the function evaluated at x. In other words, for every x-value in the domain of the function, we find the corresponding y-value by applying the function, and we mark the point (x, y) on the plane. By connecting these points, we form a visual representation of the function, allowing us to see how the output of the function changes as the input changes.

For instance, let's consider a simple function $f(x) = x$. The graph of this function is a straight line passing through the origin, making a 45-degree angle with the x-axis. This line represents all the points where the y-value equals the x-value, as dictated by our function.

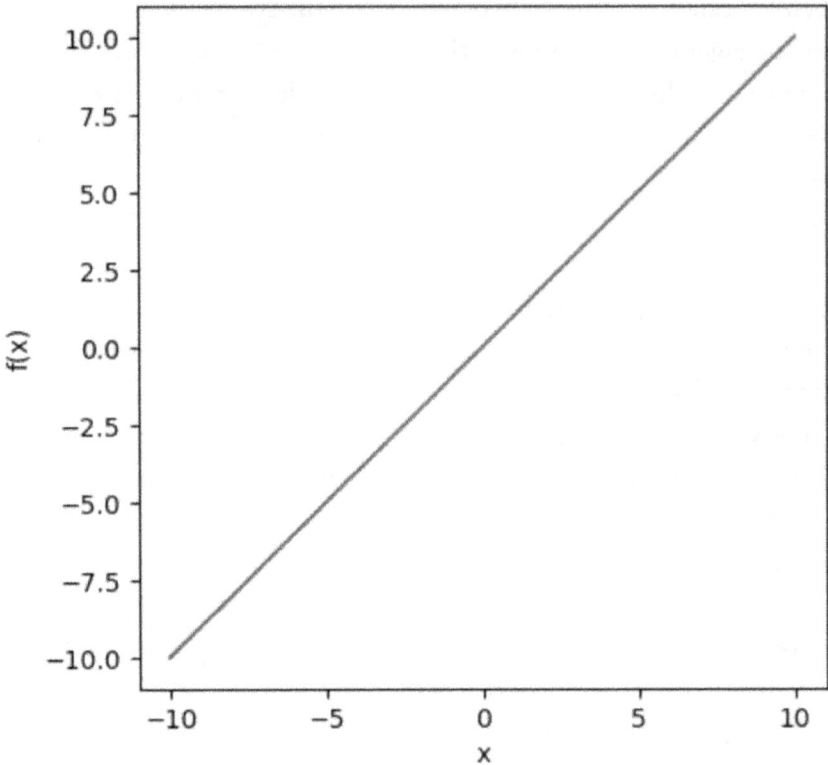

Figure 2: Graph of f(x) = x

The ability to graph functions gives us insight into many characteristics of the function. For example, where does the function reach its maximum or minimum values? Where does the function intersect the x-axis? The points of intersection are called 'roots' or 'zeros' of the function and represent the x-values where the function value is zero, which in the case of Figure 2 is 0.

Graphing also illuminates the concept of 'slope', which represents the rate of change of a function and is fundamentally tied to the concept of a derivative in calculus. For a linear function, the slope is constant, and it's represented by the angle of the line. For more complex functions, the slope can change from point to point, giving the graph its shape.

Being able to read and interpret these features on a graph is an essential skill in calculus. As we further delve into the subject, we will use graphs extensively to understand and illustrate concepts, as well as to solve problems. Thus, familiarity with the Cartesian coordinate system and the process of graphing functions will serve as a solid foundation for your study of calculus.

1.4. Intro to Limits and Continuity

As we tread deeper into the realms of calculus, two concepts stand out as particularly crucial: limits and continuity. They form the bedrock upon which much of calculus is built, laying the groundwork for understanding derivative and integral calculus.

Let's start with the concept of a 'limit'. Imagine you're approaching a traffic light. As you get closer and closer to the traffic light, your distance to it decreases, approaching zero. In calculus, we capture this notion of 'getting arbitrarily close' with the concept of a limit.

Specifically, we say that the limit of a function $f(x)$ as x approaches a certain value (let's call it 'a') is the value that $f(x)$ gets arbitrarily close to as x gets arbitrarily close to 'a'. We denote this as lim (x->a) $f(x)$ = L, where L is the value that $f(x)$ gets close to.

Now, it's important to understand that 'x getting arbitrarily close to a' is not the same as 'x is equal to a'. It's like saying you're getting very close to the traffic light, but not necessarily reaching it. This makes the concept of limit extremely useful in dealing with situations where a function is undefined or behaves erratically at a particular point.

For instance, consider the function $f(x) = x/x$. For all x ≠ 0, $f(x)$ equals 1. But what happens at x=0? The function is undefined because we're dividing by zero. But if we consider the limit as x approaches 0, we can see that $f(x)$ gets arbitrarily close to 1, even though it's undefined at x=0. Thus, we would say that lim (x->0) x/x = 1.

Next, let's delve into 'continuity', a concept intimately related to limits. We say a function $f(x)$ is continuous at a point 'a' if the value of the function at 'a' is equal to the limit of the function as x approaches 'a'. In other words, if lim (x->a) f(x) = f(a), then the function is continuous at 'a'.

Continuity gives us a way to talk about functions that don't have any 'breaks' or 'jumps' at a given point. For example, the function f(x) = x^2 is continuous everywhere. No matter what value of x you pick, the value of the function at that point is the same as the limit of the function as x approaches that point.

Understanding limits and continuity is crucial to calculus. The concept of a limit allows us to deal with values that a function approaches but doesn't necessarily reach, giving us a tool to handle infinity and infinitesimals, key concepts in calculus. Continuity, on the other hand, lets us understand the behavior of functions at specific points and across intervals, which is essential in both differential and integral calculus.

As we move forward, these concepts will become even more important. They're foundational to understanding the derivative (the rate at which a quantity changes) and the integral (the accumulation of quantities), the two central ideas in calculus. In the upcoming chapters, we'll explore these ideas in more detail, revealing the true power and beauty of calculus.

1.5. Exercises and Problem-Solving Techniques

At this point, we have familiarized ourselves with the preliminary concepts necessary to begin studying calculus. We have delved into the ideas of mathematical functions, revisited necessary algebraic principles, touched on coordinate systems and graphing, and dipped our toes into the concepts of limits and continuity. It is now time to put this knowledge to the test and hone our problem-solving skills.

One of the primary ways to practice the material we have covered so far is by working through exercises. First, though, I'll provide a framework for designing and solving problems based on what we've covered, so you can create and solve your own. This will not only test your understanding but also build the critical skill of problem formulation.

First, let's consider problems related to mathematical functions. You could create a function, such as a polynomial or a trigonometric function, and ask questions about its domain, range, roots, and graph. This could involve finding the values of the function at specific points, identifying zeros, or sketching the function's graph.

For instance, consider the function $f(x) = x^2 - 2x - 3$. Questions might include, "What is the value of $f(x)$ when $x = 3$?" or "What are the roots of the function?" or "What is the domain and range of the function?"

When graphing functions, you might draw the function and ask questions about its behavior. For example, if you graph the function $f(x) = \sin(x)$, you could ask, "What are the x-values for which $f(x) = 0$?" or "What is the maximum value of the function, and at what x-values does it occur?"

Next, for limits and continuity, design problems where you must compute the limit of a function as it approaches a specific value. These problems could also involve finding where a function is continuous or discontinuous.

For example, if you consider the function $f(x) = (x^2 - 4)/(x - 2)$, you could ask, "What is the limit of the function as x approaches 2?" Even though the function is undefined at $x = 2$, the limit exists and can be found.

When approaching these problems, remember the importance of persistence and patience. While it can sometimes be challenging to find a solution, the process of struggling with a problem can often lead to a deeper understanding of the material.

Moreover, learning to check your work is crucial. Whenever possible, use different methods to solve the same problem and see if you get the same answer. Or use your answer to make a prediction and then check whether that prediction is true.

Remember, the goal of these exercises is not just to find the right answer, but to understand why that answer is correct. By designing and solving your own problems, you'll be actively engaging with the material, which is one of the best ways to learn.

As we move forward in our study of calculus, we'll encounter increasingly complex problems that will require us to apply these concepts in innovative ways. But for now, take the time to practice and become comfortable with these fundamentals, as they form the foundation upon which the rest of calculus is built.

Exercise 1: Consider the function $f(x) = 2x + 5$. Determine the value of $f(x)$ when $x = 3$.

Exercise 2: Find the roots of the function $g(x) = x^2 - 9$.

Exercise 3: Compute the limit of the function $h(x) = (x^2 - 4)/(x - 2)$ as x approaches.

Solutions

Exercise 1: Consider the function $f(x) = 2x + 5$. Determine the value of $f(x)$ when $x = 3$.

Solution: To find the value of $f(x)$ when $x = 3$, substitute the value of x into the function: $f(3) = 2(3) + 5 = 6 + 5 = 11$. Therefore, $f(3) = 11$.

Exercise 2: Find the roots of the function $g(x) = x^2 - 9$.

Solution: To find the roots, set $g(x)$ equal to zero and solve for x: $x^2 - 9 = 0$. Factoring the expression gives $(x - 3)(x + 3) = 0$. This equation is satisfied when $x - 3 = 0$ or $x + 3 = 0$. Therefore, the roots of the function are $x = 3$ and $x = -3$.

Exercise 3: Compute the limit of the function $h(x) = (x^2 - 4)/(x - 2)$ as x approaches.

Solution: To find the limit, substitute the value of x into the function and simplify: $\lim (x\text{->}2) \, h(x) = \lim (x\text{->}2) \, [(x^2 - 4)/(x - 2)]$.

Plugging in $x = 2$ directly results in an undefined expression (division by zero).

However, by factoring the numerator, we can rewrite the expression as: $\lim (x\text{->}2) \, [(x + 2)(x - 2)/(x - 2)]$.

Cancelling out the common factor of $(x - 2)$, we have: $\lim (x\text{->}2) \, (x + 2) = 4$.

Therefore, the limit of $h(x)$ as x approaches 2 is 4.

CHAPTER 2

Understanding Limits

2.1. Defining Limits: An Intuitive Approach

With a solid grounding in the basics, it's time to delve deeper into one of the fundamental concepts in calculus: the limit. We've already encountered limits in the previous chapter, but here, we'll start unpacking the idea in a more detailed and rigorous manner. We'll begin with an intuitive approach, allowing us to grasp the concept on a more instinctive level before diving into its formal definition.

The concept of a limit involves considering what happens to a function as its input (or variable) approaches a specific value. It's like asking, "As I get closer and closer to a certain point, what does my function value get close to?"

For instance, let's consider a simple function, $f(x) = x$. The limit of $f(x)$ as x approaches 2 is simply 2, because as we take values of x that are closer and closer to 2, $f(x) = x$ gets closer and closer to 2.

The strength of the concept of a limit is that it allows us to deal with situations where we cannot directly substitute the value of x into the function. This can happen if the function is undefined at that point or if it exhibits erratic behavior.

To illustrate, let's take the function $f(x) = (x^2 - 1)/(x - 1)$. If we try to substitute $x = 1$, we find that the function is undefined. However, if we take values of x that are very close to 1, we find that $f(x)$ gets very close to 2. So, we say that the limit of f(x) as x approaches 1 is 2, denoted as lim (x->1) f(x) = 2.

Another critical point to understand about limits is that they consider the behavior of the function from both sides of the given point. That is, we look at what happens as x approaches from values less than the point (the left-hand limit) and from values greater than the point (the right-hand limit). For a limit to exist, the function must approach the same value from both sides.

For instance, consider the function f(x) = 1/(x - 2). As x approaches 2 from values less than 2 (from the left), the function values decrease without bound, going toward negative infinity. As x approaches 2 from values greater than 2 (from the right), the function values increase without bound, going towards positive infinity. Because the function does not approach the same value from both sides, we say the limit as x approaches 2 does not exist.

This intuitive approach provides a mental image for understanding limits. It helps us grasp the idea that a limit is about the value that a function approaches, not necessarily the value it reaches. This understanding is crucial as we move forward into more complex topics in calculus, including derivatives and integrals, which fundamentally rely on the concept of limits.

In the next section, we'll build on this intuitive understanding to explore the more formal definition of limits and learn how to calculate limits using algebraic methods. This will allow us to handle more complex and abstract functions and situations, further deepening our mastery of calculus.

2.2. Rules of Limits

Understanding limits intuitively is our first step into the world of calculus. Now, we will dive into the rules of limits which will provide us with a set of tools to handle them more formally and perform calculations involving limits. These rules are essentially shortcuts that allow us to calculate limits without having to rely on intuition or graphical representations, making it possible to compute limits for a wider range of functions.

Our first rule is the *Constant Rule*. This rule states that the limit of a constant as x approaches any value is simply the constant itself. In other words, if c is a constant, then lim (x->a) c = c for any value of a. This is because a constant does not change its value, no matter what x is or what it's approaching.

Next is the *Identity Rule*. According to this rule, the limit of x as x approaches a certain value 'a' is simply 'a'. Formally, this can be stated as lim (x->a) x = a. This rule makes sense intuitively as the closer x gets to 'a', the closer the value of x (which is our function in this case) will be to 'a'.

Then, we have the *Sum/Difference Rule*. This rule tells us that the limit of the sum or difference of two functions is equal to the sum or difference of their respective limits. If we have two functions f(x) and g(x), we can state this rule as lim (x->a) [f(x) ± g(x)] = lim (x->a) f(x) ± lim (x->a) g(x), assuming the two individual limits on the right side exist.

The *Product Rule*, as you might guess, deals with the multiplication of functions. It states that the limit of the product of two functions is the product of their individual limits. Formally, this is expressed as lim (x->a) [f(x) * g(x)] = lim (x->a) f(x) * lim (x->a) g(x), again assuming the two individual limits on the right side exist.

Lastly, we have the *Quotient Rule*. This rule states that the limit of the quotient of two functions is the quotient of their limits, provided that the limit of the denominator is not zero (as division by zero is undefined). This can be stated as lim (x->a) [f(x) / g(x)] = lim (x->a) f(x) / lim (x->a) g(x), as long as lim (x->a) g(x) ≠ 0.

These rules greatly simplify the process of finding limits and apply to a wide variety of functions. They make it possible to break complex expressions down into simpler parts, compute the limits of those parts, and then reassemble the results.

It's crucial to remember that these rules apply only when the individual limits exist. In some cases, you might encounter indeterminate forms or other situations where these rules don't apply. Such situations require additional techniques, such as factoring, rationalizing, or *L'Hopital's Rule*, which we will explore in later sections.

Understanding these rules will make you more proficient at handling limits, a foundational skill in calculus. In the next sections, we will build upon this foundation to delve into the concept of derivatives, one of the two main branches of calculus, where the importance of limits becomes more evident.

2.3. Special Limits

As we delve deeper into the world of calculus, we'll encounter situations that require more than just the standard rules of limits. These are the cases when we face indeterminate forms, where traditional algebraic manipulations aren't enough to find the limit. Two common forms that we will consider in this section are $0/0$ and ∞/∞. To deal with these situations, we need to employ special limit techniques.

First, consider the indeterminate form $0/0$. This form arises when both the numerator and denominator of a fraction tend towards zero as the variable approaches a certain value. A common example is the limit of the function $f(x) = (\sin x) / x$ as x approaches 0. In such a case, we cannot directly substitute $x = 0$ into the function because it results in the form $0/0$, which is indeterminate. However, by applying L'Hopital's Rule—a method that uses derivatives to evaluate limits—we find that this limit equals 1. L'Hopital's Rule will also be covered in a later chapter as it builds on the concept of derivatives.

Another situation we often face is the indeterminate form ∞/∞. This form arises when both the numerator and denominator of a function tend towards infinity as the variable approaches a certain value. A classic example is the limit of the function $f(x) = (x^2) / x$ as x approaches infinity. In this case, we cannot directly substitute infinity into the function because it results in the indeterminate form ∞/∞. However, with a little algebraic manipulation—specifically, by dividing the numerator and denominator by x, the highest power of x in the denominator—we can simplify the function to $f(x) = x$, for which the limit as x approaches infinity is, naturally, infinity.

Finally, there are special limits involving exponential and logarithmic functions. For example, the limit of $(1 + 1/x)^x$ as x approaches infinity is the number 'e' (~2.71828), the base of natural logarithms. This result is remarkable because it connects the idea of limits with exponential growth, and forms the basis for continuous compounding in financial mathematics.

In each of these examples, we see the power of limits in handling intricate situations. They require a blend of algebraic manipulation, intuitive understanding, and sometimes additional techniques, like L'Hopital's Rule, to evaluate.

As we further explore the landscape of calculus, these special limits will be invaluable tools, enabling us to tackle complex problems with confidence.

2.4. Limits at Infinity

Having considered limits at finite values, it's time to explore the concept of limits at infinity. This concept answers questions like "What happens to the value of our function as our input grows very large or decreases without bound?" In this context, infinity is not a number but a symbol that represents an unbounded quantity, something that is growing without any limit.

When we say the limit of a function $f(x)$ as x approaches positive infinity (denoted as $\lim (x\text{->}\infty) f(x)$) or negative infinity (denoted as $\lim (x\text{->} -\infty) f(x)$), we are seeking the value that the function approaches as x gets arbitrarily large or small. Sometimes, the function may approach a specific number. Other times, the function may continue to grow or decrease without bound.

Consider a simple function, $f(x) = x$. As x grows larger and larger (approaches ∞), the value of $f(x)$ also grows larger without bound. So, we say the limit as x approaches ∞ is ∞, written as $\lim (x\text{->}\infty) f(x) = \infty$.

Now, consider another function, $g(x) = 1/x$. As x grows larger (approaches ∞), the value of $g(x)$ gets smaller and smaller, getting infinitely close to 0. So, we say the limit as x approaches ∞ is 0, written as lim (x->∞) g(x) = 0. Similarly, as x becomes more negative (approaches -∞), g(x) also approaches 0, so lim (x-> -∞) g(x) = 0.

As another example, take the function h(x) = 2x^2 - 3x + 1. As x approaches ∞, the term with the highest power of x, in this case, 2x^2, dominates the behavior of the function. The values of -3x and 1 become increasingly insignificant as x grows larger, so the limit as x approaches ∞ is ∞, written as lim (x->∞) h(x) = ∞. As x approaches -∞, 2x^2 still dominates, but because it's being multiplied by 2, the limit as x approaches -∞ is also ∞, written as lim (x-> -∞) h(x) = ∞.

These examples illustrate that the concept of a limit at infinity can be thought of as capturing the long-term behavior of a function as the variable moves towards positive or negative infinity. In some cases, we may see that functions tend towards a certain number (as in g(x) = 1/x), while in others, they continue to increase or decrease without bound (as in f(x) = x and h(x) = 2x^2 - 3x + 1).

Limits at infinity are a powerful tool in calculus, offering insights into the global or large-scale behavior of functions. They lay the foundation for concepts like horizontal asymptotes, and they play a significant role in the study of series and improper integrals, topics we will explore in later chapters. Understanding limits at infinity, therefore, opens the door to deeper and more nuanced applications of calculus.

2.5. Exercises and Problem-Solving Techniques

With a solid understanding of limits, both at finite values and infinity, it's now time to apply this knowledge through exercises and develop a variety of problem-solving techniques. These exercises will span the entire gamut of this chapter's content, from simple limit evaluations to more complex applications, including special limits and limits at infinity.

A critical aspect of learning calculus is consistent practice, so let's walk through a problem-solving approach for dealing with limit problems.

Consider the following example problem: Find the limit as x approaches 2 for the function $f(x) = (x^2 - 4) / (x - 2)$.

Step 1: Substitute the value into the function

The first step is to substitute $x = 2$ into the function. However, doing so results in a 0/0 indeterminate form. This means we'll need to use a different method.

Step 2: Simplify the function

Since substitution didn't work, let's try simplifying the function. We notice that the numerator can be factored using the difference of squares: $x^2 - 4 = (x - 2)(x + 2)$. This simplification cancels out the x - 2 in the numerator and the denominator, resulting in a simpler function: $f(x) = x + 2$.

Step 3: Substitute the value into the simplified function

Now, if we substitute $x = 2$ into the simplified function, we get $f(2) = 2 + 2 = 4$. Thus, the limit as x approaches 2 for the function is 4.

Each problem you encounter might require different techniques, such as factoring, rationalizing the numerator, or dividing by the highest power of x. Keep in mind that in some cases a limit might not exist if the function approaches different values from the left and right, or if it oscillates without settling at a specific value.

As you practice, you'll become more comfortable with these techniques and start to recognize when each one is most useful. Remember that practice is key to mastering calculus. Take the time to work through many different types of problems, always reflecting on your solution process to continually improve your problem-solving skills.

In the following chapters, we will build on our understanding of limits to introduce the concept of the derivative, which is one of the two fundamental concepts in calculus. As we delve deeper, you'll see that all these concepts are interconnected in a beautiful and intricate web, revealing the elegant structure that underlies calculus.

CHAPTER 3

Introduction to Derivatives

3.1. What is a Derivative?

Now that we have a firm understanding of limits, it's time to use this knowledge to explore one of the two central concepts in calculus: the derivative.

In its essence, a derivative measures how a function changes as its input changes. In more concrete terms, it measures the rate at which a quantity is changing at a given point. This concept is instrumental in numerous fields of science and engineering, from physics, where it describes velocity and acceleration, to economics, where it measures rates of change in various economic variables.

To help visualize this, think about a car trip. If you consider the car's position as a function of time, the derivative of this function at any given moment would provide the car's velocity, or speed, at that specific instant.

Let's now formally define the derivative of a function.

Given a function f(x), the derivative of f at a specific point x = a, denoted as f'(a), is defined as: **f'(a) = lim (h->0) [(f(a + h) - f(a)) / h]** where h is a very small number.

This formula essentially measures the rate of change of the function f(x) at the point x = a. If this limit exists, we say that the function f(x) is differentiable at the point x = a.

Here, the term (f(a + h) - f(a)) represents the change in the function's output (or 'rise') over a tiny interval h, while h itself is the change in input (or 'run').

So, the whole expression [(f(a + h) - f(a)) / h] can be seen as the 'rise over run', which is the slope of the line tangent to the function at point x = a. Therefore, we can also think of the derivative as providing the slope of the function at a given point.

For example, let's consider the function **f(x) = x^2**. Using the limit definition of the derivative, we can calculate the derivative at x=a as follows:

f'(a) = lim (h->0) [(f(a + h) - f(a)) / h]
Substituting f(x) = x^2:
= lim (h->0) [((a + h)^2 - a^2) / h]
Expanding (a + h)^2:
= lim (h->0) [(a^2 + 2ah + h^2 - a^2) / h]
Canceling out the a^2 terms:
= lim (h->0) [(2ah + h^2) / h]
Simplifying by dividing each term by h:
= lim (h->0) [2a + h]
Taking the limit as h approaches 0:
= 2a

Therefore, the derivative of f(x) = x^2 at any point a is equal to 2a.

So, the derivative of f(x) = x^2 at any point x = a is 2a. This means that the rate of change of the function x^2 at any point x = a is 2a, and the slope of the tangent line to the function at this point is also 2a.

Understanding the concept of the derivative is crucial for navigating the world of calculus. It provides a window into the behavior of functions, offering a detailed snapshot of how they change at each point. With this introduction to derivatives, we can now move on to understanding the dynamics of change, rates, and motion.

3.2. Physical and Geometrical Interpretation of Derivatives

The concept of the derivative, while loaded with mathematical symbols and notation, has profound physical and geometric interpretations. Let's delve deeper into how derivatives relate to the world around us and the shapes we plot on our graphs.

Firstly, let's consider the physical interpretation of a derivative. Suppose we have a function that represents an object's position in space with respect to time, denoted as $s(t)$. If we compute the derivative of $s(t)$, written as $s'(t)$ or ds/dt, we get a new function that represents the velocity (the rate at which something changes its position) of the object at any given instant of time. In this context, the derivative is measuring the rate of change of position, i.e., how fast the position is changing per unit of time. This is essentially the speed of the object at each moment. If we further differentiate the velocity function, we obtain the acceleration of the object, which measures the rate of change of velocity.

Imagine driving a car on a straight road. At the start, your car is stationary, and as you press the gas pedal, your car begins to move and gain speed. The speedometer shows your velocity, but it does not tell you how that speed changes as you press or release the gas pedal—this is where the derivative comes in. If you were to graph your speed over time, the slope at each point (the rate of change of your speed) would be your acceleration, indicating how rapidly you are speeding up or slowing down.

Now, let's shift our attention to the geometric interpretation. Here, the derivative gives us the slope of the tangent line to the function at any given point. Remember, the tangent line is the line that just touches the function at that point and nowhere else in an infinitesimally small interval around that point.

Consider a roller coaster ride. The track of the roller coaster is analogous to the graph of a function, and the tangent line at each point on the track corresponds to the derivative at that point. At the peaks and valleys where the coaster changes direction, the tangent line (and hence the derivative) is zero, indicating a momentary pause in the upward or downward movement. On the other hand, the steepest parts of the track correspond to the points where the derivative is at its maximum or minimum values, indicating the most rapid ascent or descent.

By looking at the derivative of a function, we can get a detailed sense of the function's behavior. The places where the derivative is zero, for example, correspond to local maxima or minima of the function, or places where the function levels out. Understanding these characteristics is crucial when studying the behavior of functions and can provide significant insights in fields such as physics, economics, biology, and more.

In essence, derivatives help us quantify change, providing a precise mathematical language for describing rates of change and the shape of functions. They form a bridge between abstract mathematical functions and tangible, physical phenomena, making calculus an indispensable tool in understanding the world around us.

3.3. Rules for Derivatives

As we begin to work more with derivatives, you'll be relieved to know that there are established rules that make finding the derivative of a function more straightforward than using the limit definition every time. These rules, derived from the limit definition, allow us to compute derivatives more quickly and efficiently.

Let's delve into some of these crucial derivative rules:

Constant Rule: The derivative of a constant is zero. This is intuitive because a constant doesn't change, and the derivative measures the rate of change. Therefore, if c is a constant, the derivative, denoted as d/dx(c), equals zero.

Power Rule: The derivative of x raised to the power of n, denoted as $d/dx(x^n)$, equals $n*x^{(n-1)}$. So, for example, the derivative of x^3 would be $3x^2$. This rule is especially useful for polynomials and allows us to find their derivatives term by term.

Product Rule: This rule is used when dealing with the product of two functions. The derivative of the product of two functions u(x) and v(x) is not merely the product of their derivatives. Instead, it's given by: d/dx(uv) = u'v + uv', where u' and v' are the derivatives of u and v, respectively.

Quotient Rule: Similar to the product rule, the quotient rule is used when finding the derivative of a quotient of two functions. The derivative of u(x) divided by v(x) is given by: **d/dx(u/v) = (vu' - uv') / v^2**, where u' and v' are the derivatives of u and v, respectively.

Chain Rule: The chain rule is used when we have a composite function, a function inside another function, say f(g(x)). The chain rule states that the derivative of a composite function is the derivative of the outer function evaluated at the inner function, multiplied by the derivative of the inner function. It's denoted as: **d/dx(f(g(x))) = f'(g(x)) * g'(x)**.

It's crucial to note that these rules are not standalone but can often be used together when faced with more complex functions. Knowing when to apply which rule often depends on recognizing the function's structure and having enough practice with different types of problems.

Having a robust understanding of these derivative rules will save considerable time and effort, making it easier and more efficient to handle derivatives. As you continue to explore the world of calculus, these rules will prove to be invaluable tools for analysis and problem-solving. In the subsequent sections, we will apply these rules to find derivatives of various types of functions, further solidifying your understanding of these key mathematical tools.

3.4. Applications of Derivatives

Derivatives, with their potent ability to gauge rates of change, are immensely useful in a wide array of applications across science, engineering, economics, and even social sciences. In this section, we will examine a few key areas where derivatives play a significant role, helping us understand and predict real-world phenomena.

Physics and engineering: One of the most significant applications of derivatives is in the field of physics, where they give us the tools to describe motion, forces, and energy. For example, the derivative of a position function with respect to time gives the velocity and the derivative of velocity yields acceleration. These principles underpin the whole study of kinematics. Engineers use derivatives to optimize designs, perform structural analysis, control systems, and much more.

Economics and business: Derivatives are a central tool in economics, allowing economists to model and predict various economic variables' behavior. For example, the derivative of a cost function gives the marginal cost, or the rate of change of the cost with respect to the number of items produced. Similarly, the derivative of a revenue function gives the marginal revenue, the rate of change of the revenue when the quantity sold changes. These insights can help businesses optimize their operations and maximize profit.

Biology and medicine: In the life sciences, derivatives can be used to model population dynamics, enzyme reactions, drug dosage and clearance, and much more. For example, the rate of change of a population (the derivative of the population size) can be modeled as proportional to the current population size, leading to exponential growth or decay.

Computer Science and data analysis: In machine learning and data analysis, derivatives play a vital role in optimization algorithms, such as gradient descent, which are used to find the best model parameters to fit the data. The derivative helps guide the algorithm toward the lowest error, optimizing the model's predictive ability.

Environmental science: Derivatives help model and predict the behavior of environmental variables over time, such as the concentration of a pollutant in a body of water or the rate of change of global temperatures.

What all these applications have in common is the idea of change—how one quantity changes in relation to another and how we can use that information to understand, predict, and sometimes control that change. As we continue to explore calculus, we'll see even more ways in which the concepts of derivatives and integration, the other pillar of calculus, can help us make sense of the world.

Understanding how derivatives apply in these various contexts will not only deepen your appreciation for calculus but will also give you the practical tools to apply these mathematical concepts to real-world problems. As you continue to delve into the world of calculus, keep in mind these applications of derivatives and how they connect abstract mathematical concepts to the real world.

3.5. Exercises and Problem-Solving Techniques

Below are 10 sample questions and solutions based on the concepts discussed in this chapter.

Exercise 1: Find the derivative of the function $f(x) = 5x^3 - 2x^2 + 7x - 4$.

Exercise 2: Calculate the derivative of the function $g(x) = 4e^x - 2\ln(x)$.

Exercise 3: Determine the derivative of the function $h(x) = (2x^2 - 3x + 1)/(x^2 + 1)$.

Exercise 4: Given the function $f(x) = \sin(3x)$, find its derivative.

Exercise 5: Find the derivative of the function $g(x) = \sqrt{x} - x^2 + 5x^3$.

Exercise 6: Calculate the derivative of the function $h(x) = 4x^3 - 2x^2 + 7x - 4\sqrt{x}$.

Exercise 7: Find the derivative of the function $f(x) = e^{(2x)} - \ln(3x)$.

Exercise 8: Determine the derivative of the function $g(x) = (x^2 + 3x)/(x^3 - 2)$.

Exercise 9: Given the function $h(x) = \cos(4x)$, find its derivative.

Exercise 10: Find the derivative of the function $f(x) = 2\sqrt{x} - 3x^2 + 4/x$.

Solutions

Exercise 1: Find the derivative of the function $f(x) = 5x^3 - 2x^2 + 7x - 4$.
Solution: To find the derivative of the given function, we apply the power rule. The derivative of each term is obtained by multiplying the coefficient by the power of x and reducing the power by 1. Therefore, the derivative of $f(x)$ is $f'(x) = 15x^2 - 4x + 7$.

Exercise 2: Calculate the derivative of the function $g(x) = 4e^x - 2\ln(x)$.
Solution: To find the derivative of $g(x)$, we use the rules for differentiating exponential and logarithmic functions. The derivative is $g'(x) = 4e^x - 2/x$.

Exercise 3: Determine the derivative of the function $h(x) = (2x^2 - 3x + 1)/(x^2 + 1)$.
Solution: To find the derivative of $h(x)$, we apply the quotient rule. The derivative is given by $h'(x) = [(2x^2 + 2) - (4x^2 - 6x + 2)] / (x^2 + 1)^2 = (-2x^2 + 6x) / (x^2 + 1)^2$.

Exercise 4: Given the function $f(x) = \sin(3x)$, find its derivative.
Solution: To find the derivative of $f(x)$, we use the chain rule. The derivative is $f'(x) = 3\cos(3x)$.

Exercise 5: Find the derivative of the function $g(x) = \sqrt{x} - x^2 + 5x^3$.
Solution: To find the derivative of $g(x)$, we differentiate each term separately. The derivative is $g'(x) = (1/2)x^{-1/2} - 2x + 15x^2$.

Exercise 6: Calculate the derivative of the function $h(x) = 4x^3 - 2x^2 + 7x - 4\sqrt{x}$.
Solution: To find the derivative of $h(x)$, we differentiate each term using the power rule and the chain rule. The derivative is $h'(x) = 12x^2 - 4x + 7 - 2/(2\sqrt{x}) = 12x^2 - 4x + 7 - 1/\sqrt{x}$.

Exercise 7: Find the derivative of the function $f(x) = e^{(2x)} - \ln(3x)$.

Solution: To find the derivative of f(x), we differentiate each term using the rules for exponential and logarithmic functions. The derivative is $f'(x) = 2e^{(2x)} - 1/x$.

Exercise 8: Determine the derivative of the function $g(x) = (x^2 + 3x)/(x^3 - 2)$.
Solution: To find the derivative of $g(x)$, we apply the quotient rule. The derivative is given by $g'(x) = [(2x + 3)(x^3 - 2) - (x^2 + 3x)(3x^2)] / (x^3 - 2)^2$.

Exercise 9: Given the function $h(x) = \cos(4x)$, find its derivative.
Solution: To find the derivative of $h(x)$, we use the chain rule. The derivative is $h'(x) = -4\sin(4x)$.

Exercise 10: Find the derivative of the function $f(x) = 2\sqrt{x} - 3x^2 + 4/x$.
Solution: To find the derivative of $f(x)$, we differentiate each term using the power rule and the quotient rule. The derivative is $f'(x) = (1/\sqrt{x}) - 6x - 4/x^2$.

CHAPTER 4

Understanding the Concept of Integration

4.1. Chain Rule

The *chain rule* is an invaluable tool in calculus, especially when dealing with composite functions. A composite function is a function that consists of two (or more) functions. This is best illustrated by an example, such as $f(g(x))$, where the function g is applied first and then the function f.

But how do we differentiate a composite function? Well, this is where the chain rule comes in. The chain rule states that the derivative of a composite function is the derivative of the outer function, evaluated at the inner function, and multiplied by the derivative of the inner function. If we have a composite function $f(g(x))$, we can denote its derivative as $(f(g(x)))' = f'(g(x)) * g'(x)$.

To understand why this is, consider the composite function as a process in two stages. The inner function, $g(x)$, changes the input x, and then the outer function, $f(u)$, changes the $g(x)$ result. When we take the derivative, we are considering how small changes in the input x affect the final result after both stages. The chain rule takes into account both the rate of change of the outer function with respect to $g(x)$ $(f'(g(x)))$ and the rate of change of the inner function with respect to x $(g'(x))$.

In practice, applying the chain rule involves the following steps:

Identify the inner and outer functions: The inner function is the one that is "inside" the other in the composite function, while the outer function is the one that is applied to the result of the inner function.

Differentiate the outer function: Apply your knowledge of basic derivative rules to find the derivative of the outer function, treating the inner function as a single variable.

Evaluate the derivative of the outer function at the inner function: Substitute the inner function into the derivative of the outer function. This gives you the derivative of the outer function evaluated at the inner function.

Differentiate the inner function: Again, apply basic derivative rules to find the derivative of the inner function.

Multiply the results: The final step in applying the chain rule is to multiply the derivative of the outer function (evaluated at the inner function) by the derivative of the inner function. This gives you the derivative of the composite function.

The chain rule is a powerful tool in calculus, providing a method to differentiate complex functions that are composed of simpler parts. Understanding and applying the chain rule requires practice, so in the subsequent sections, we will work through various examples and exercises to solidify your understanding of this vital calculus rule.

4.2. Product Rule and Quotient Rule

In calculus, sometimes the function we need to differentiate is a product or quotient of two functions. To handle these cases efficiently, we employ two key rules: the *product rule* and the *quotient rule*.

Let's start with the product rule. Suppose we have two functions, $u(x)$ and $v(x)$, and we're interested in finding the derivative of their product. According to the product rule, the derivative of $u(x)v(x)$ is given by $u'(x)v(x) + u(x)v'(x)$. This equation means that the derivative of the product of two functions is the derivative of the first function multiplied by the second function, added to the first function multiplied by the derivative of the second function.

Now, why does this rule work? Well, the essence of differentiation is finding the rate of change. When you change the input x a little, both u(x) and v(x) change. The product **u(x)v(x)** changes because u changes and because v changes, so we need to take both changes into account, which is what the product rule does.

Moving on to the quotient rule. The quotient rule is used when the function is a quotient, i.e., a division of two functions, say **u(x)/v(x)**. The quotient rule states that the derivative of **u(x) / v(x)** is given by **[v(x)u'(x) – u(x)v'(x)] / [v(x)]²**.

The interpretation of the quotient rule is slightly more complex than the product rule. The derivative of the quotient of two functions is the bottom function times the derivative of the top function minus the top function times the derivative of the bottom function, all divided by the square of the bottom function. The square in the denominator reflects the fact that when we change the input x, the bottom function v(x) changes, and this changes not just the fraction itself (as would be the case for the product) but also the "size" or scale of the fraction, since the denominator determines how large the fraction is relative to 1.

For this reason, both the product and quotient rule are essential tools in calculus. They allow us to break down complex expressions into simpler parts, making differentiation a more manageable task.

4.3. Implicit Differentiation

In previous sections, we mostly dealt with functions written in explicit form, where y is expressed as a function of x, such as $y = f(x)$. However, in many mathematical and real-world scenarios, we encounter equations where y and x are intermingled in a way that it's hard or impossible to solve for y explicitly. This is where *implicit differentiation* comes in handy.

Implicit differentiation is a powerful technique in calculus used to find the derivative of equations not easily solved for y, or equations involving variables implicitly. In these situations, it might be extremely cumbersome or even impossible to explicitly isolate y before taking the derivative. Implicit differentiation provides a way around this.

Implicit differentiation allows us to find the derivative of y with respect to x, denoted **dy/dx or y′**, without having to solve the equation explicitly for y. Instead, we take advantage of the fact that y is a function of x (even if not stated directly) and use the chain rule to differentiate.

Let's delve into how implicit differentiation works with a simple example. Consider the equation of a circle $x^2 + y^2 = r^2$, where r is the radius. To find dy/dx, or the rate of change of y with respect to x, we differentiate both sides of the equation with respect to x.

When we differentiate x^2 with respect to x, we get 2x, which is straightforward. However, when we differentiate y^2 with respect to x, we use the chain rule from earlier in this chapter, which states that the derivative of a composite function is the derivative of the outer function, evaluated at the inner function, multiplied by the derivative of the inner function. Here, our "outer" function is $f(u) = u^2$, and the "inner" function is $g(x) = y$. Using the chain rule, the derivative of y^2 with respect to x is **2y * (dy/dx)**.

So the derivative of the entire equation $x^2 + y^2 = r^2$ is 2x + 2y * (dy/dx) = 0, since the derivative of a constant (r^2 in this case) is 0. Now, we can solve for dy/dx to find **dy/dx = -x/y**. This gives us the slope of the tangent line to the circle at any point (x, y).

What if you have a more complex equation, such as **y = x*y²** or **y = sin(xy)**? In these cases, each term containing a y would require the use of the chain rule. The key is to carefully apply the chain rule at each step, remembering to include the dy/dx each time you differentiate a y term.

In summary, implicit differentiation is a versatile tool that expands the types of equations we can handle. It allows us to find derivatives without the need to solve the equation for y explicitly. This technique also becomes useful for dealing with complex equations in higher math, physics, economics, or numerous other fields.

4.4. Higher Order Derivatives
The beauty of calculus is that it allows us to look not just at the rate of change of a function, but also at how that rate of change itself

changes. The first derivative of a function gives us the function's rate of change or slope at a given point, but we don't have to stop there. By taking the derivative of a derivative, we can get what's called a "second derivative", which tells us how the slope of a function is changing. If we continue this process, taking the derivative of the second derivative, we get the "third derivative", and so on. These are known as higher order derivatives.

The second derivative is particularly important. While the first derivative at a certain point describes the slope of the tangent line at that point, the second derivative describes the concavity of the function at that point. In other words, it tells us whether the function is curving upwards or downwards.

If the second derivative is positive at a certain point, it means the first derivative is increasing at that point, so the function is becoming steeper as you move from left to right. We call this kind of shape "concave up". If the second derivative is negative, the function is "concave down". If the second derivative is zero, the function could be changing from concave up to concave down or vice versa, or it could be a point of inflection.

A point of inflection is a point on the curve of the function where it changes concavity. In other words, it's where the curve switches from curving upwards to curving downwards, or vice versa. The second derivative can help us locate these points. However, just like with first derivatives, setting the second derivative equal to zero only gives us candidates for points of inflection. We need to test these points to see if the concavity actually changes.

To understand why concavity and points of inflection are important, consider the physical meaning of the second derivative in the context of motion. If the function $f(x)$ represents the position of an object, then the first derivative $f'(x)$ represents velocity, or the rate of change of position. The second derivative $f''(x)$, being the rate of change of velocity, represents acceleration. So, if an object is accelerating, it means the velocity is increasing, and the distance-time graph is concave up. If the object is decelerating, the graph is concave down.

The process of taking higher order derivatives continues in a similar manner. The third derivative, or the derivative of the acceleration function in the above example, would represent the rate of change of acceleration, also known as *jerk*. The fourth derivative, referred to as *jounce*, represents the rate of change of the jerk, and so on. However, in practice, most applications don't go beyond the second derivative.

Higher order derivatives provide us with deeper insights into the behavior of functions, allowing us to visualize and understand complex changes. Mastering the interpretation and calculation of these derivatives is vital for further study in calculus, as well as many application areas in physics, economics, and engineering.

4.5. Exercises and Problem-Solving Techniques

Exercise 1: Find the derivative of the function $f(x) = \sin(2x) + \cos(3x)$.

Exercise 2: Calculate the derivative of the function $g(x) = \ln(4x^2 + 5x)$.

Exercise 3: Differentiate the function $h(x) = e^x * \cos(x)$.

Exercise 4: Find the derivative of the function $f(x) = (2x + 1)^4$.

Exercise 5: Calculate the derivative of the function $g(x) = \text{sqrt}(3x^2 + 2x + 1)$.

Exercise 6: Differentiate the function $h(x) = x^3 * e^x$.

Exercise 7: Find the derivative of the function $f(x) = 1 / (2x + 3)$.

Exercise 8: Calculate the derivative of the function $g(x) = \ln(x^2 + 1) / x$.

Exercise 9: Differentiate the function $h(x) = (\sin(x))^2 + (\cos(x))^2$.

Exercise 10: Find the second derivative of the function $f(x) = 5x^3 - 2x^2 + 4x$.

Solutions

Exercise 1: Find the derivative of the function $f(x) = \sin(2x) + \cos(3x)$.
Solution: To find the derivative of f(x), we differentiate each term using the chain rule. The derivative is $f'(x) = 2\cos(2x) - 3\sin(3x)$.

Exercise 2: Calculate the derivative of the function $g(x) = \ln(4x^2 + 5x)$.
Solution: To differentiate g(x), we apply the chain rule. The derivative is $g'(x) = (8x + 5) / (4x^2 + 5x)$.

Exercise 3: Differentiate the function $h(x) = e^x * \cos(x)$.
Solution: To differentiate h(x), we use the product rule. The derivative is $h'(x) = e^x * \cos(x) - e^x * \sin(x)$.

Exercise 4: Find the derivative of the function $f(x) = (2x + 1)^4$.
Solution: To differentiate f(x), we apply the chain rule. The derivative is $f'(x) = 4(2x + 1)^3 * 2$.

Exercise 5: Calculate the derivative of the function $g(x) = \sqrt{3x^2 + 2x + 1}$.
Solution: To differentiate g(x), we use the chain rule. The derivative is $g'(x) = (6x + 2) / (2\sqrt{3x^2 + 2x + 1})$.

Exercise 6: Differentiate the function $h(x) = x^3 * e^x$.
Solution: To differentiate h(x), we use the product rule. The derivative is $h'(x) = 3x^2 * e^x + x^3 * e^x$.

Exercise 7: Find the derivative of the function $f(x) = 1 / (2x + 3)$.
Solution: To differentiate f(x), we use the quotient rule. The derivative is $f'(x) = -2 / (2x + 3)^2$.

Exercise 8: Calculate the derivative of the function $g(x) = \ln(x^2 + 1) / x$.
Solution: To differentiate g(x), we use the quotient rule and the chain rule. The derivative is $g'(x) = (2x / (x^2 + 1) - \ln(x^2 + 1)) / x^2$.

Exercise 9: Differentiate the function h(x) = (sin(x))^2 + (cos(x))^2.

Solution: To differentiate h(x), we use the chain rule and the fact that (sin(x))^2 + (cos(x))^2 = 1. The derivative is h'(x) = 0.

Exercise 10: Find the second derivative of the function f(x) = 5x^3 - 2x^2 + 4x.

Solution: To find the second derivative of f(x), we first differentiate f(x) to obtain f'(x) = 15x^2 - 4x + 4. Then, we differentiate f'(x) to find the second derivative f''(x) = 30x - 4.

CHAPTER 5

Techniques of Integration

5.1. What is Integration?

In this section, we will work through various exercises that involve the concepts we've discussed in this chapter: the chain rule, product rule, quotient rule, implicit differentiation, and higher order derivatives. These exercises are designed to strengthen your understanding of these concepts and to hone your problem-solving techniques in calculus. Let's now delve into the details.

Applying the Chain Rule

The chain rule is most useful when dealing with composite functions, where one function is nested inside another. Consider the function $h(x) = (5x^2 - 3x + 7)^{\wedge}4$. To find its derivative, we must apply the chain rule.

First, let's identify the outer and inner functions. Here, the outer function is $g(u) = u^{\wedge}4$ and the inner function is $f(x) = 5x^2 - 3x + 7$. Applying the chain rule, we get: $h'(x) = g'(f(x)) * f'(x)$. That's the derivative of the outer function, evaluated at the inner function, multiplied by the derivative of the inner function. Solving, we get $h'(x) = 4(5x^2 - 3x + 7)^3 * (10x - 3)$.

The Power of the Product and Quotient Rules

When faced with the product or quotient of two functions, the product and quotient rules are indispensable.

Let's find the derivative of the function $p(x) = (x^2 + 1)(\sin(x))$. We can see that this is a product of two functions, $f(x) = x^2 + 1$ and $g(x) = \sin(x)$.

Applying the product rule, we get $p'(x) = f'(x)g(x) + f(x)g'(x)$. Calculating these derivatives, we find $p'(x) = (2x)(\sin(x)) + (x^2 + 1)(\cos(x))$.

Navigating Implicit Differentiation

Implicit differentiation is vital when dealing with equations where it's impractical or impossible to solve for y explicitly. Let's consider the equation of an ellipse $x^2/4 + y^2/9 = 1$. Applying implicit differentiation, we differentiate both sides of the equation with respect to x, and solve for dy/dx, which gives us the slope of the tangent line at any point (x, y) on the ellipse.

Exploring Higher Order Derivatives

Consider the function $f(x) = x^5 - 3x^3 + 2x$. Its first derivative is $f'(x) = 5x^4 - 9x^2 + 2$. If we take the derivative of $f'(x)$, we get the second derivative, $f''(x) = 20x^3 - 18x$. Now, we can continue this process to obtain the third derivative, fourth derivative, and so on.

These exercises offer a blend of the practical applications of these rules and techniques. However, understanding and proficiency will not come overnight. Calculus is a subject that requires continuous practice and an active attempt to understand the concepts at play. It's not enough to simply understand how to take derivatives—you should understand what these derivatives mean, how they're used, and why they're important.

5.2. Fundamental Theorem of Calculus

At the heart of calculus lies a powerful, pivotal concept known as the *Fundamental Theorem of Calculus*. This theorem establishes the profound relationship between the two central operations in calculus: differentiation and integration.

The Fundamental Theorem of Calculus is split into two parts:

Part 1: This part concerns the computation of definite integrals. If a function 'f' is continuous over the interval [a, b] and 'F' is an antiderivative of 'f' on [a, b], then the definite integral from 'a' to 'b' of 'f' with respect to 'x' equals 'F(b) - F(a)'. That is, ∫ from a to b of **f(x) dx = F(b) − F(a)**.

This part simplifies the process of evaluating definite integrals. Instead of calculating the limit of a Riemann sum, which can be quite laborious, we find an antiderivative of the function we are integrating and evaluate it at the limits of integration.

For example, to find the definite integral from 1 to 3 of the function **f(x) = x²**, we first find an antiderivative **F(x) = (1/3)x³**, and then calculate F(3) – F(1), which equals 8.

Part 2: This part illustrates that differentiation and integration are reverse processes. If 'f' is a function that is continuous over the interval [a, b], then the function 'F' defined by F(x) = ∫ from a to x of f(t) dt for all x in [a, b], is continuous on [a, b] and differentiable on (a, b), and F'(x) = f(x) for all x in (a, b).

Part 2 tells us that if we take a function 'f(x)', integrate it, and then differentiate the result, we end up back at the original function 'f(x)'. This underpins many aspects of calculus, unifying it into a coherent field of study.

Both parts of the theorem have a vast range of practical applications. For instance, in physics, the first part of the theorem is often used to determine the total distance traveled given the speed function, while the second part is essential for solving differential equations, which have countless applications in fields ranging from engineering to biology.

Understanding the Fundamental Theorem of Calculus is crucial not only because it is the central theorem of calculus, but also because it facilitates the understanding of the relationship between integrals and derivatives, which are the foundational building blocks of calculus. It is through this theorem that we can truly appreciate the beauty and depth of calculus, recognizing how seemingly disparate concepts are intricately intertwined.

5.3. Indefinite and Definite Integrals

Integration, one of the two main operations of calculus, can be somewhat perplexing but it can be distilled into two types: indefinite and definite integrals.

Indefinite Integral:

Also known as the antiderivative, the indefinite integral of a function 'f(x)' represents a family of functions, each of which can be differentiated to obtain the original function. That is, if 'F(x)' is the antiderivative of 'f(x)', then the derivative of 'F(x)' equals 'f(x)'.

Let's elaborate on this with an example: the indefinite integral of $f(x) = x^2$ is $F(x) = (1/3)x^3 + C$. Note the "+ C" at the end. This represents the constant of integration and it indicates that there is a whole family of functions that are antiderivatives of $f(x) = x^2$. For instance, $(1/3)x^3 + 1$, $(1/3)x^3 - 5$, and $(1/3)x^3$ are all antiderivatives of $f(x) = x^2$.

Definite Integral:

The definite integral, on the other hand, can be thought of as the limit of a sum, or as an accumulated total. It is associated with a specific interval from 'a' to 'b' and is denoted as ∫ from a to b of f(x) dx.

The definite integral represents the signed area between the x-axis and the curve represented by the function f(x) from x = a to x = b. If f(x) is positive over this interval, then this area lies above the x-axis and is positive. If f(x) is negative, the area is below the x-axis and is considered negative. If the function crosses the x-axis, the total area is the sum of the absolute values of the negative and positive areas.

Consider the function $f(x) = x^2$ on the interval [1, 3]. We can compute the definite integral using the antiderivative $F(x) = (1/3)x^3$. The Fundamental Theorem of Calculus, Part 1, tells us that the definite integral equals $F(3) - F(1) = 9 - 1/3 = 8$ and $2/3$.

It's important to remember that while the indefinite integral represents a family of functions, the definite integral, since it depends on specific values of x (a and b), is a number.

Understanding both indefinite and definite integrals paves the way for more complex topics such as integration techniques and applications, which include finding areas, volumes, work, and solutions to differential equations. Integration, whether indefinite or definite, is an indispensable tool in the mathematics toolbox, and it plays a vital role in diverse scientific and engineering fields.

5.4. Applications of Integration

Integration, a fundamental operation in calculus, goes far beyond the classroom's theoretical exercises. It has profound applications across diverse fields, including physics, engineering, economics, and statistics, among others.

Physics and engineering: In physics, integration is often used to find quantities that involve a rate of change over an interval. Consider the example of a moving object with a variable velocity. If we want to find the total distance traveled over a time period, we can integrate the velocity function over that time period. This is a direct application of the definite integral, as it represents the accumulation of a quantity (distance) over a specific interval.

Similarly, in electrical engineering, the voltage across a capacitor or an inductor in an electrical circuit can be determined by integrating the current flowing through the circuit over time. This demonstrates how integration aids in the analysis and design of complex electrical systems.

Economics: In economics, integration is used in a variety of contexts. One common application is in the computation of consumer and producer surplus, which represents the monetary gain obtained by consumers and producers by participating in the market. The surplus can be computed as the area under the demand (or supply) curve but above (or below) the price level, which requires the use of definite integration.

Statistics: In statistics, integration plays a vital role in the concept of probability density functions (PDFs). A PDF is a function that describes the likelihood of a random variable taking on a particular value. The total probability of all possible outcomes is equal to one, which is found by integrating the PDF over all possible values.

Environmental science: In environmental science, integration is employed to model population growth, particularly in scenarios where the growth rate is not constant but depends on factors such as the current population, available resources, or time. The solution to these models involves integrating a differential equation.

In summary, these applications underscore the utility and relevance of integration in various professional and academic fields. The fundamental concepts and techniques in calculus, such as differentiation and integration, serve as robust tools for problem-solving and quantitative analysis, enhancing our understanding and interaction with the world around us.

5.5. Exercises and Problem-Solving Techniques

An advantage of calculus, particularly of integration, is not only its theoretical elegance but also its practical use as a tool for solving real-world problems. To get better at integration, consistent practice and the application of problem-solving techniques are key. In this section, we will walk through a variety of exercises designed to challenge your understanding of integration, from indefinite and definite integrals to applications in diverse contexts.

Exercise 1: Evaluating Indefinite Integrals

Our first exercise is a simple task to find the indefinite integral of a given function. The function is $f(x) = 3x^2 - 2x + 1$. Remember, when you find the antiderivative, don't forget to add the constant of integration, C.

As a solution guide, recall the basic power rule for integration: $\int x^n \, dx = (1/(n+1))x^{(n+1)} + C$. Applying this rule to each term in the function should yield the antiderivative.

Exercise 2: Evaluating Definite Integrals

For the second exercise, let's evaluate a definite integral. Let's consider the function $g(x) = x^2$ on the interval [1, 3]. Remember, you need to find the antiderivative of g(x) and then apply the Fundamental The-orem of Calculus, Part 2, which involves subtracting the antiderivative evaluated at the lower limit of the interval from the antiderivative evalu-ated at the upper limit.

Exercise 3: Physical Application of Integration

For this exercise, consider a car that accelerates at a rate of **a(t) = 6t m/s² for 0 ≤ t ≤ 10 seconds**. Find the total distance the car travels in these 10 seconds.

Hint: You'll need to integrate the acceleration function to find the velocity function, and then integrate the velocity function to find the position function.

Exercise 4: Economic Application of Integration

Suppose the demand curve for a product in a market is given by the function **d(p) = 1000 – 20p**, where p is the price. The market price is set at $25. Compute the consumer surplus in this market.

Hint: You'll need to integrate the demand function from 0 to 25 and then subtract the rectangle area under the price level.

Exercise 5: Environmental Science Application of Integration

Consider a bacteria population that grows at a rate of **P'(t) = 0.5P**, where P is the population size, and t is the time in hours. If initially there are 100 bacteria, find the population size after 2 hours.

Hint: This is a simple differential equation which can be solved by integrating both sides appropriately.

As you work through these exercises, remember that problem-solving in calculus is about understanding the principles behind the operations and the context of the problems. It is not merely about computation; it's also about interpretation and application.

Solutions

Exercise 1: Evaluating Indefinite Integrals
Solution: To find the indefinite integral of f(x), we apply the power rule for integration to each term separately. The power rule states that the integral of x^n dx is equal to $(1/(n+1))x^{\wedge}(n+1)$ + C, where C is the constant of integration.

$\int (3x^2 - 2x + 1) \, dx = (3/3)x^3 - (2/2)x^2 + x + C$

$= x^3 - x^2 + x + C$

So, the indefinite integral of f(x) is $x^3 - x^2 + x + C$.

Exercise 2: Evaluating Definite Integrals
Solution: To evaluate the definite integral, we first find the antiderivative of x^2, which is $(1/3)x^3$. Then, we apply the Fundamental Theorem of Calculus, Part 2, which states that the definite integral of a function f(x) from a to b is equal to F(b) − F(a), where F(x) is the antiderivative of f(x).

$\int[1, 3] \, x^2 \, dx = [(1/3)x^3]$ [from 1 to 3]

$= [(1/3)(3)^3] - [(1/3)(1)^3]$

$= (1/3)(27) - (1/3)(1)$

$= 9 - 1$

$= 8$

So, the value of the definite integral is 8.

Exercise 3: Physical Application of Integration
Solution: To find the total distance traveled by the car, we need to integrate the velocity function. Since the acceleration is given, we integrate it to obtain the velocity function. Then, we integrate the velocity function to find the position function.

First, integrate the acceleration function to find the velocity function:

$v(t) = \int(6t) \, dt$

$= 3t^2 + C_1$

Next, integrate the velocity function to find the position function:

$s(t) = \int(3t^2 + C_1) \, dt$

$= t^3 + C_1 t + C_2$

Now, we can find the total distance traveled by evaluating the position function at t = 10 and subtracting the position at t = 0:

Distance = s(10) – s(0)

$= (10^3 + C_1(10) + C_2) - (0^3 + C_1(0) + C_2)$

$= 1000 + 10C_1 + C_2 - 0 - 0 - C_2$

$= 1000 + 10C_1$

So, the total distance traveled by the car in 10 seconds is $1000 + 10C_1$ meters.

Exercise 4: Economic Application of Integration

Solution: To compute the consumer surplus, we need to find the area between the demand curve and the price level of $25. This area represents the extra benefit consumers receive by paying a price lower than their maximum willingness to pay.

First, we integrate the demand function from 0 to 25 to find the total benefit:

Benefit = ∫[0, 25] (1000 – 20p) dp

$= [1000p - 10p^2]$ [from 0 to 25]

$= (1000(25) - 10(25)^2) - (1000(0) - 10(0)^2)$

= 25000 – 6250 – 0

= 18750

Next, we find the area of the rectangle under the price level:

Area = 25 * 1000 = 25000

Finally, we compute the consumer surplus by subtracting the area of the rectangle from the total benefit:

Consumer Surplus = Benefit – Area

= 18750 – 25000

= -6250

The negative value indicates that the consumer surplus is zero in this market, as consumers are paying exactly their maximum willingness to pay.

Exercise 5: Environmental Science Application of Integration

Solution: To find the population size after 2 hours, we need to solve the differential equation by integrating both sides appropriately.

Separate the variables by dividing both sides by P: (1/P) dP = 0.5 dt

Integrate both sides:

∫(1/P) dP = ∫0.5 dt

ln|P| = 0.5t + C

Now, solve for P by exponentiating both sides:

|P| = e^(0.5t + C)

P = ± e^(0.5t + C)

Given that initially, there are 100 bacteria (P = 100) at t = 0, we can determine the value of the constant C:

100 = ± e^(0.5(0) + C)

100 = ± e^C

$C = \ln(100)$

Substituting $C = \ln(100)$ into the equation, we get:

$P = \pm\, e\verb|^|(0.5t + \ln(100))$

$P = \pm\, 100e\verb|^|(0.5t)$

Since the population size cannot be negative, we take the positive solution:

$P = 100e\verb|^|(0.5t)$

Now, substitute $t = 2$ into the equation to find the population size after 2 hours:

$P(2) = 100e\verb|^|(0.5(2))$

$P(2) = 100e\verb|^|1$

$P(2) \approx 271.83$

Therefore, the population size after 2 hours is approximately 271.83 bacteria.

CHAPTER 6

Techniques of Integration

6.1. Substitution Method

In Chapter 5, we discussed basic integration techniques and their applications, but some integrals are not easy to handle with the basic rules of integration alone. This is where the *Substitution Method*, also known as *U-Substitution*, comes into play. This method can be seen as the counterpart to the Chain Rule in differentiation. It's the first of many techniques that we will explore in this chapter to tackle more complex integrals.

To understand the idea of the Substitution Method, let's start with a straightforward example. Suppose you're asked to find the integral of a function such as $f(x) = 2x \cos(x^2)$, it's not immediately clear how we could directly apply the basic rules of integration to solve this problem. However, by using substitution, we can simplify the integral and make it easier to solve.

The general procedure for using substitution involves a few steps:

Choosing a substitution: You start by selecting a part of the integral that you'll define to be a new variable, u. In our example, it would make sense to let u be the inside of the cosine function, that is, $u = x^2$.

Computing the differential du: Once you have your substitution, you take its derivative with respect to x to get du. In this case, du would be equal to 2x dx. Notice that 2x dx is part of our original integral, which makes our choice of u = x^2 a good one.

Substituting in the integral: Now we substitute u and du into the integral, effectively transforming it into an integral in terms of u. This gives us ∫cos(u) du.

Performing the integral in terms of u: In this simpler form, the integral is often straightforward to evaluate. Here, **∫cos(u) du = sin(u) + C.**

Substituting back the original variable: Finally, we substitute the original expression for u back into the result, yielding the solution in terms of the original variable, x. In this case, sin(u) + C becomes sin(x^2) + C.

A few key points to note: Not all integrals are suitable for the Substitution Method, and sometimes it takes a bit of practice to see what the best choice for u is. Additionally, remember to always substitute back in the original variable at the end, since the problem started in terms of that variable.

In more complex scenarios, you might need to adjust the limits of integration when dealing with definite integrals, or sometimes even apply the substitution method more than once. As we progress further into calculus, this technique becomes an essential tool in your mathematical arsenal, paving the way for more advanced integration methods.

6.2. Integration by Parts

After having mastered the substitution method in the previous section, it's time to learn another powerful tool to deal with more complex integrals: the *method of integration by parts*.

This method is the integration counterpart to the Product Rule for differentiation. Just as the Product Rule provides a way to differentiate the product of two functions, integration by parts gives us a way to integrate the product of two functions. This technique is incredibly useful when trying to integrate products where one function can easily be differentiated, and the other can be easily integrated.

The rule for integration by parts is derived directly from the Product Rule and is usually given as follows:

$\int u\,dv = uv - \int v\,du$

Here, u and v are functions of the variable we're integrating with respect to. To use this rule, we must choose which part of our integrand to assign to u and which part to dv. These choices can make the difference between an easy problem and a hard one.

Let's illustrate this concept with an example. Let's consider the integral of the product of x and e^x, that is, $\int x*e^x\,dx$. For this example, we can let u = x, meaning that dv is the remaining part of the integrand, e^x dx.

Once we've made this choice, we derive du and integrate dv to get v. So du = dx, and v = $\int e^x\,dx$ = e^x.

Applying the integration by parts formula, we get:

$\int x*e^x\,dx = x * e^x - \int e^x\,dx$

The remaining integral, $\int e^x\,dx$, is much simpler and can be directly integrated to e^x.

So, our final result is:

$\int x*e^x\,dx = x * e^x - e^x + C$

It's worth mentioning that, just like in the substitution method, the choice of u and dv can be critical to simplifying the problem. A commonly used acronym to aid in making this choice is LIATE (Logarithmic, Inverse trigonometric, Algebraic, Trigonometric, Exponential), which prioritizes the choice of u based on the type of function.

Furthermore, in some cases, applying integration by parts multiple times or using it in combination with other methods like substitution might be necessary. Integration by parts is a powerful method in your integration toolkit, but it also requires practice and insight to apply effectively.

6.3. Partial Fractions

One of the most significant challenges in calculus is finding the integral of rational functions, i.e., functions that are the ratio of two polynomials. For some rational functions, we can use the simple rules of integration or even the substitution method that we just discussed. However, for more complicated rational functions, we need to deploy a technique known as the *method of partial fractions*.

The method of partial fractions is a mathematical technique used to simplify complex fractions by breaking them down into simpler fractions that are easier to handle. This technique is particularly useful when you need to integrate or differentiate a complex rational function.

To give you a sense of how the method works, let's suppose we have a fraction that is the ratio of two polynomials: $F(x) = (2x^3 + 3x^2 - x + 1) / (x^2 - 1)$. Our task is to break down this complex fraction into simpler fractions that can be integrated individually.

The first step in this process is to check the degrees of the polynomials in the numerator and denominator. If the degree of the numerator is greater than or equal to the degree of the denominator, we use polynomial division to simplify the fraction until we get a polynomial plus a fraction, where the numerator has a lower degree than the denominator.

For our example, the degree of the numerator is greater than that of the denominator, so we perform polynomial division first. However, if the degree of the numerator had been less than the denominator, we could have moved directly to decomposing into partial fractions.

The next step is to factor the denominator into its irreducible factors. In our case, $x^2 - 1$ can be factored as $(x - 1)(x + 1)$.

The partial fractions will have these factors as their denominators. The numerators will be unknown constants that we must solve for. We represent the original fraction as a sum of these new fractions: $F(x) = A/(x - 1) + B/(x + 1)$.

We then solve for the constants A and B by combining the fractions on the right-hand side back into a single fraction and equating coefficients with the original function $F(x)$.

Once we've found the values of A and B, we have successfully decomposed $F(x)$ into simpler fractions that can be integrated individually using the basic rules of integration.

It's important to note that while the method of partial fractions can initially seem challenging, it's a powerful technique that can greatly simplify the process of integrating complex rational functions. With enough practice, you will be able to easily use this method to solve a variety of integral problems. The key is to be patient and carefully follow each step of the process.

6.4. Improper Integrals

In our exploration of integration, we've mainly focused on integrals of functions over closed and bounded intervals. However, there are many practical problems that require integrating a function over an infinite interval or integrating a function that has an infinite discontinuity within the interval of integration. These scenarios involve what we call "improper integrals".

An improper integral is an integral that has either infinite limits or an integrand that approaches infinity at some points in the interval of integration. Although these integrals might seem "improper", we can still find their values using well-defined mathematical processes.

Infinite Intervals

Firstly, let's consider the case where we need to find the integral of a function over an infinite interval. For example, the integral $\int$ from 1 to ∞ of $1/x^2$ dx.

In situations like these, we cannot directly apply the fundamental theorem of calculus because it only applies to integrals over finite intervals. Instead, we introduce a parameter 'a' that stands in for infinity. So, our integral becomes $\int$ from 1 to a of $1/x^2$ dx.

We then take the limit of this integral as 'a' approaches infinity. In mathematical terms, this is written as lim (as a->∞) $\int$ from 1 to a of $1/x^2$ dx. This is known as taking the improper integral "at infinity". If this limit exists, we say that the improper integral converges. If the limit does not exist, we say that the integral diverges.

Infinite Discontinuities

The other type of improper integral involves an integrand that approaches infinity at some points within the interval of integration.

For instance, consider the integral ∫ from 0 to 1 of 1/x dx. Here, the function 1/x becomes infinite as x approaches 0.

To handle this situation, we again use the concept of limits. We replace the problematic point with a parameter 'b' and take the limit as 'b' approaches the problematic point. So, the integral ∫ from 0 to 1 of 1/x dx becomes lim (as b->0+) ∫ from b to 1 of 1/x dx.

Once again, if this limit exists, we say the improper integral converges, otherwise, it diverges.

These might seem like complex concepts but they are an essential part of calculus and many other areas of mathematics. Improper integrals open the door to solving a wide range of problems, from finding areas under infinite curves to solving differential equations. Just like with other calculus techniques, it takes practice to become proficient with improper integrals, but the effort will pay off as you expand your mathematical toolkit.

6.5. Exercises and Problem-Solving Techniques

Applying what we've learned about advanced integration techniques requires practice and experimentation. Below are several exercises designed to help solidify your understanding of these concepts. Following each problem, we will walk through how to approach solving it and the insights that can be gained from the solution.

Exercise 1: Evaluate the improper integral ∫ from 1 to ∞ of $1/x^2$ dx.

Exercise 2: Evaluate the improper integral ∫ from 0 to 1 of $1/\sqrt{x}$ dx.

Exercise 3: Evaluate the improper integral ∫ from 0 to ∞ of $e^{(-x)}$ dx.

Solutions

Exercise 1: Evaluate the improper integral ∫ from 1 to ∞ of $1/x^2$ dx.

Solution: We start by recognizing that this is an improper integral because the upper limit is infinity. To handle this, we'll replace the upper limit with a variable 'a' and take the limit as 'a' approaches infinity.

So, the integral ∫ from 1 to ∞ of $1/x^2$ dx becomes lim (as a->∞) ∫ from 1 to a of $1/x^2$ dx. We can compute the antiderivative of $1/x^2$ as $-1/x$, then evaluate it at the limits of integration.

This leads us to lim (as a->∞) [-1/a - (-1/1)], which simplifies to lim (as a->∞) (1 - 1/a).

As 'a' approaches infinity, 1/a goes to 0, so the limit is 1. Therefore, the improper integral converges and its value is 1.

Exercise 2: Evaluate the improper integral ∫ from 0 to 1 of 1/√x dx.

Solution: This improper integral has a singularity at x = 0, so we need to take the limit as the lower bound approaches 0.

∫ from 0 to 1 of 1/√x dx = lim (as b->0+) ∫ from b to 1 of 1/√x dx

To integrate 1/√x, we can rewrite it as $x^{(-1/2)}$ and use the power rule for integration:

$\int x^{(-1/2)} dx = 2x^{(1/2)} + C$

Now, we evaluate the definite integral with the limits of integration:

lim (as b->0+) [2√x - 2√b]

= lim (as b->0+) (2√x - 2√b)

= 2√1 - 2√0

= 2 - 0

= 2

So, the value of the improper integral is 2.

Exercise 3: Evaluate the improper integral ∫ from 0 to ∞ of $e^{(-x)}$ dx.

Solution: To evaluate this improper integral, we need to take the limit as the upper bound approaches infinity.

$\int$ from 0 to ∞ of $e^{\wedge}(-x) \, dx = \lim$ (as a->∞) $\int$ from 0 to a of $e^{\wedge}(-x) \, dx$
To integrate $e^{\wedge}(-x)$, we can use the rule for the integral of exponential functions:
$\int e^{\wedge}(-x) \, dx = -e^{\wedge}(-x) + C$

Now, we evaluate the definite integral with the limits of integration:
$\lim$ (as a->∞) $[(-e^{\wedge}(-a)) - (-e^{\wedge}(-0))]$
$= \lim$ (as a->∞) $(-e^{\wedge}(-a) + 1)$
$= -0 + 1$
$= 1$
So, the value of the improper integral is 1.

CHAPTER 7

Introduction to
Differential Equations

7.1: What is a Differential Equation?

A differential equation is a type of equation that relates a function with one or more of its derivatives. The term "differential" comes from the word "difference", relating to change, hence these are equations that describe how things change over time or space according to some rule.

To comprehend what a differential equation is, it is essential to understand the term "derivative". In calculus, a derivative measures how a function changes as its input changes. In simple terms, it's the rate of change or the slope of the function at a certain point.

In a differential equation, the function we are trying to find, often denoted as 'y' or 'f(x)', is not known. What is known, however, is the relationship between the function and its derivative(s). In this context, the 'solution' to a differential equation is the function or functions that satisfy the equation.

There are several types of differential equations, categorized based on several factors including the type of derivatives involved, the order of the highest derivative, the linearity of the equation, and whether they are ordinary or partial differential equations.

Ordinary Differential Equations (ODEs): These involve functions of one variable and their derivatives. An example of an ODE is the equation $dy/dx = x$, where dy/dx is the derivative of y with respect to x.

Partial Differential Equations (PDEs): These involve functions of multiple variables and their partial derivatives. An example of a PDE is the heat equation, $\partial u/\partial t = \partial^2 u/\partial x^2$, where $\partial u/\partial t$ and $\partial^2 u/\partial x^2$ are the partial derivatives of u with respect to time t and position x, respectively.

Linear Differential Equations: These are equations in which the dependent variable and its derivatives are to the first power and are not multiplied together.

Nonlinear Differential Equations: These can contain functions of the dependent variable and its derivatives raised to a power, or the dependent variable and its derivatives multiplied together.

Homogeneous Differential Equations: These are equations in which every term is a function of the dependent variable and its derivatives. There are no terms involving only the independent variable or constants.

Nonhomogeneous Differential Equations: These include terms involving the independent variable or constants.

The order of a differential equation is determined by the highest order derivative involved in the equation. For example, a second order differential equation involves the second derivative of a function. The equation $d^2y/dx^2 = x$ is a second order ODE.

Differential equations are pervasive in various fields of science and engineering. They are used to describe numerous physical and mathematical systems, including the motion of planets (Newton's Second Law), population dynamics, electrical circuits, fluid dynamics, and heat conduction, to name just a few.

In the subsequent sections of this chapter, we will delve into the techniques for solving different types of differential equations and explore their applications. It's important to note that while some differential equations can be solved exactly, others may only have approximate solutions, or might not have a known solution at all.

7.2. Solving First Order Differential Equations

In our journey into the realm of differential equations, we first need to acquaint ourselves with *first-order differential equations*. These equations involve the first derivative of a function, and perhaps the function itself, but not any higher derivatives.

A first-order differential equation is typically expressed as follows: The derivative of y with respect to x equals a function f that depends on x and y.

In mathematical terms, we'd say **dy/dx = f(x, y)**. When we talk about solving such an equation, we're looking for a function y that depends on x, written as y(x), which makes the equation hold true.

Several methods exist for solving first-order differential equations. Let's delve into the two most common types: separable equations and linear equations.

Separable Equations

Firstly, a first-order separable equation is one where the right side, our function f(x, y), can be expressed as a product of a function of x and a function of y. We could, for example, write it as **dy/dx = g(x) multiplied by h(y).** If we can rewrite the equation in this form, we say the differential equation can be "separated" into two integrals. Essentially, we rewrite the equation as follows: The integral of dy divided by h(y) equals the integral of g(x) with respect to x.

Then, we integrate both sides, allowing us to solve for y(x).

Linear Equations

The second type, a first-order linear differential equation, takes on the following form: The derivative of y with respect to x plus a function p(x) multiplied by y equals another function g(x). Here, both p(x) and g(x) are functions of x.

The technique to solve such equations involves a procedure known as using integration factors. Though we won't go into the details here, the general solution to such an equation is as follows:

y(x) is equal to the exponential function of negative integral of p(x)dx, all multiplied by the integral of the exponential function of integral of p(x)dx multiplied by g(x)dx plus a constant of integration C.

While these methods offer powerful tools for solving first-order differential equations, not all such equations can be solved exactly. In such instances, we may resort to numerical methods. As broad and rich as this field of study is, these examples just scratch the surface of what's possible with differential equations.

7.3. Applications of Differential Equations

The strength of differential equations lies in their broad applicability. In the real world, many phenomena, from natural science to economics, can be modeled using differential equations. Let's delve into a few examples.

Physics: Newton's Law of Cooling

In physics, one of the most classic applications of differential equations is modeling Newton's Law of Cooling, which describes the rate of change of the temperature of an object over time. According to the law, the rate at which an object cools (i.e., its temperature decreases) is proportional to the difference between its temperature and the ambient temperature. This law can be modeled as a first-order differential equation: $dT/dt = -k(T - T_a)$, where T is the temperature of the object, T_a is the ambient temperature, k is the cooling constant, and t is the time.

Biology: The Logistic Growth Model

Differential equations also find extensive use in biology, where they model population dynamics. One such model is the Logistic Growth Model, a first-order differential equation that describes how a population (P) changes with time (t) in an environment with limited resources: $dP/dt = rP(1 - P/K)$. Here, r is the maximum per-capita growth rate, and K is the carrying capacity, or the maximum population size the environment can support.

Economics: The Malthusian Growth Model

In economics, differential equations model various phenomena. The Malthusian Growth Model, for instance, predicts population growth in the absence of resource limitations. It can be represented by a simple first-order differential equation: $\mathbf{dP/dt = rP}$, where P is the population, r is the growth rate, and t is time.

Engineering: Electric Circuits

In electrical engineering, the behavior of circuits, especially those involving capacitors and inductors, is typically described using differential equations. The voltage across a charging or discharging capacitor over time, for example, can be represented by a first-order differential equation derived from Kirchhoff's voltage law.

While we've highlighted a few domains, the real-world applications of differential equations span far and wide. From the microscopic behavior of particles to the large-scale structure of the universe, from the rhythm of the heart to the spread of diseases, differential equations are fundamental to capturing the essence of the world around us. Their study is not just a theoretical exercise; it's a journey into understanding the complex dynamics that govern our universe.

7.4. Exercises and Problem-Solving Techniques

Let's cement the understanding you have developed in this chapter with some practice exercises and problem-solving techniques related to differential equations. We'll tackle problems concerning real-world applications, bringing theory to life.

Exercise 1: Cooling a cup of coffee

Suppose you're cooling a cup of coffee in a room where the temperature is a constant 25 degrees Celsius. The coffee is initially 80 degrees Celsius, and after 10 minutes, it cools to 60 degrees Celsius. Model this scenario with Newton's Law of Cooling, and find how long it will take for the coffee to cool to 30 degrees Celsius.

Exercise 2: Predicting population growth

Consider a population of bacteria in a petri dish. Suppose the initial population is 500, and after 2 hours, the population is observed to have grown to 1500. The dish can support up to 5000 bacteria. Using the Logistic Growth Model, predict when the bacteria population will reach 4000.

Solutions

Exercise 1: Cooling a cup of coffee

To solve this problem, we'll first set up the differential equation as per Newton's Law of Cooling: $dT/dt = -k(T - T_a)$. We are given that T_a, the ambient temperature, is 25 degrees Celsius. T is the temperature of the coffee at a given time.

The cooling constant, k, can be found by plugging in the values from a known point in time. We know that after 10 minutes (which we'll treat as 1/6 hours to match the units of k), the temperature is 60 degrees. This gives us one equation we can solve to find k.

Once we have the value of k, we can use the differential equation to solve for T when the coffee has cooled to 30 degrees.

Exercise 2: Predicting population growth

The Logistic Growth Model is given by the differential equation $dP/dt = rP(1 - P/K)$, where P is the population, r is the growth rate, and K is the carrying capacity. Here, K is 5000.

We can find r by using the known population size after 2 hours. This will provide us with one equation in which the only unknown is r.

Once we've determined r, we can use the differential equation to solve for the time t when the population P becomes 4000.

The exercises and approaches mentioned above are a way for us to apply our knowledge of differential equations to real-world scenarios. These exercises should give you a good understanding of how to translate a physical situation into a mathematical model and then solve it. Remember, the key to mastering differential equations, like any mathematical concept, is practice. Always aim to understand the principles deeply and solve a variety of problems to become proficient.

CHAPTER 8

Sequences and Series

8.1. Understanding Sequences

As we move forward in our calculus journey, let's introduce a new topic: sequences. A sequence is a set of numbers arranged in a particular order, with each number associated with a unique positive integer called its index or term number. In other words, a sequence is a function whose domain is the set of positive integers.

To understand sequences, we should first be familiar with some key terminology.

Term: Each number in a sequence is called a term. In the sequence {1, 2, 3, 4, 5}, each of the numbers (1, 2, 3, 4, 5) are terms of the sequence.

Position/Index: This refers to the place of each term in the sequence. For instance, in the sequence {1, 2, 3, 4, 5}, the number 3 is at the 3rd position or index.

Infinite sequence: A sequence that continues indefinitely. For instance, the sequence of all natural numbers {1, 2, 3, 4, 5, ...} is an infinite sequence as it continues indefinitely.

Finite sequence: A sequence that ends after a certain number of terms is known as a finite sequence. For example, the sequence {1, 2, 3, 4, 5} is a finite sequence as it ends after 5 terms.

Real sequence: A sequence whose range is the set of real numbers is known as a real sequence. An example would be {1, 1.5, 2, 2.5, 3}.

Sequences are typically denoted as {a_n}, where 'a' represents the term and 'n' is the position of the term. For example, in the sequence {1, 2, 3, 4, 5}, a_1 = 1, a_2 = 2, a_3 = 3, and so on. The general term a_n is also known as the nth term of the sequence.

There are also different types of sequences based on the relationship between successive terms.

Arithmetic Sequence: A sequence is said to be arithmetic if the difference between any two successive terms is constant. The constant difference is also known as the common difference.

Geometric Sequence: A sequence is geometric if the ratio of any two successive terms is constant. This constant ratio is also known as the common ratio.

Understanding sequences is a vital step before we proceed to the concept of series. The idea of sequences provides the bedrock for many essential concepts in calculus and other branches of mathematics, such as number theory, real analysis, and combinatorics. It's an initial step towards comprehending the behavior of functions as their inputs become arbitrarily large, a concept that's integral to calculus.

8.2. Introduction to Series

Building upon our understanding of sequences, we now delve into a concept closely related to sequences: the concept of a series. While a sequence is a set of terms, a series is the sum of the terms of a sequence. For instance, if we have an arithmetic sequence {1, 2, 3, 4, 5}, the corresponding series would be the sum of these terms: 1 + 2 + 3 + 4 + 5.

In notation, we use the Greek letter sigma, Σ, to represent the sum of a sequence's terms. If we have a sequence a_n, where n ranges from 1 to N, then the corresponding series S_N is represented as:

S_N = Σ (from n=1 to N) a_n

S_N here represents the Nth partial sum of the sequence.

Two main types of series exist: *finite series* and *infinite series*.

Finite series: A finite series is the sum of a finite sequence's terms. Like the finite sequence, a finite series has a specific number of terms.

Infinite series: The sum of the terms of an infinite sequence is an infinite series. These are more complex, as they don't end. We often discuss the series 'converging' to a specific value.

The idea of convergence is vital in understanding series. An infinite series is said to converge if the sequence of its partial sums (S_1, S_2, S_3, ...) approaches a certain number as the number of terms goes to infinity. If it does not, we say the series diverges. The concept of limits, which we've explored in earlier chapters, is a crucial tool in determining convergence or divergence of a series.

Like sequences, series can also be classified based on the nature of their terms.

Arithmetic series: The sum of the terms of an arithmetic sequence forms an arithmetic series.

Geometric series: A series is geometric if it corresponds to a geometric sequence. Geometric series have especially useful properties that allow for easy evaluation and manipulation.

In later sections, we will explore tests to determine if an infinite series converges or diverges and delve into power series, a crucial type of series used in many areas of mathematics and physics. The understanding of series forms a cornerstone of integral calculus and provides a basis for many theorems and applications.

Remember, series extend the concept of addition: while ordinary addition allows us to add a finite number of numbers, series provide a framework to add infinitely many numbers, a key step in the journey of calculus.

8.3. Convergence and Divergence

Having introduced the concept of series, we now turn our attention to the critical questions of convergence and divergence. These terms refer to the behavior of a series as the number of terms tends to infinity.

When we talk about an infinite series 'converging,' we mean that as we add more and more terms, the partial sums of the series approach a certain finite value. For example, the geometric series $1/2 + 1/4 + 1/8 + 1/16 + \ldots$ converges to 1, because each additional term brings the total sum closer to 1.

However, not all series behave this way. Consider the series $1 + 2 + 3 + 4 + \ldots$ Here, as we add more terms, the sum just grows larger and larger without bound. We say that such a series 'diverges.'

The critical challenge of working with infinite series in calculus lies in determining whether a given series converges or diverges. If it converges, we also want to know what value it converges to. This is far from trivial, and over the centuries mathematicians have developed various tests to help answer these questions. These tests analyze the properties of the series' terms to draw conclusions about the overall behavior of the series.

Next are some key concepts we'll explore in relation to convergence and divergence.

Limit of a sequence of partial sums: For a series to converge, the sequence of its partial sums must have a limit. We denote this as the limit as n approaches infinity of S_n, where S_n is the nth partial sum of the series.

Limit of a series term: If the individual terms of a series do not approach zero as n approaches infinity, the series must diverge. This is known as the *Test for Divergence*. However, be aware that the converse is not true; if the terms do approach zero, the series may still either converge or diverge.

Comparison Tests: These tests involve comparing a given series with a series that is already known to converge or diverge.

Absolute and conditional convergence: If a series converges when all its terms are replaced by their absolute values, it is said to converge absolutely. If it converges only when the terms keep their original signs, it converges conditionally.

In subsequent sections, we will go into more depth on specific tests for convergence, and look at how these can be applied to a range of different series. This topic requires careful attention, as a thorough understanding of convergence and divergence is essential for many of the later concepts in calculus.

8.4. Power Series and Taylor Series

In the previous sections, we have examined sequences and series in a general sense. Now, we will focus on particular kinds of series that are central to the field of calculus: *power series* and *Taylor series*.

A power series is an infinite series of the form $\Sigma(a\_n*(x-c)^n)$ from n=0 to infinity. Here, x is the variable, c is a constant known as the center of the series, and the coefficients a_n form a sequence of real or complex numbers. Power series are very important in calculus and in the wider field of mathematical analysis because they can be used to represent a wide variety of functions in a form that is often more convenient for analysis and computation.

Power series have a specific interval of convergence, (r1, r2), within which the series converges to a finite value. The behavior of the series outside this interval depends on the specific series: it may converge at one or both endpoints of the interval, or it may diverge. Determining the interval of convergence is an essential part of working with power series.

Moving forward, we will find a specific type of power series that bears significant importance, called the Taylor series. Named after British mathematician Brook Taylor, a Taylor series is a representation of a function as an infinite sum of terms that are calculated from the values of the function's derivatives at a single point.

In more simple terms, Taylor series can be thought of as polynomial approximations of functions. The nth degree Taylor polynomial of a function approximates the function within a certain range by a polynomial of degree n. The more terms we include in the polynomial, the closer it comes to the actual function, and in the limit as n tends to infinity, the Taylor series of a function is equal to the function itself (within its interval of convergence).

A function's Taylor series is centered at a particular point c (which can be any real number), and is given by the formula:

$f(x) = \Sigma(f^\wedge n(c) * (x - c)^\wedge n/n!)$ from n=0 to infinity.

Here, $f^\wedge n(c)$ denotes the nth derivative of the function at the point c.

By using power series and Taylor series, we can solve complex calculus problems by working with polynomial functions, which are much easier to handle. For example, the functions sin(x), cos(x), e^x, and many others can be represented exactly by a Taylor series over the entire number line.

8.5. Exercises and Problem-Solving Techniques

In this section, we delve into exercises and problem-solving techniques associated with the concepts of sequences, series, power series, and Taylor series. Through these exercises, we aim to foster an intuitive understanding of these concepts and their applications in calculus.

Exercise 1: Find the power series representation for the function $f(x) = e^\wedge x$ centered at c = 0.

Exercise 2: Determine the interval of convergence for the power series $\Sigma((x-3)^\wedge n/n)$ from n=1 to infinity.

Solutions

Exercise 1: Find the power series representation for the function $f(x) = e^x$ centered at $c = 0$.

Solution: We begin by noting that the Taylor series for a function $f(x)$ centered at c is given by $\Sigma(f^n(c) * (x - c)^n/n!)$ from n=0 to infinity. In this case, we have $f(x) = e^x$ and $c = 0$, and we know that the derivative of e^x is simply e^x for all n. Therefore, all the terms of the series are equal to 1, and we have:

$f(x) = \Sigma(x^n/n!)$ from n=0 to infinity.

This is the power series representation for e^x centered at $c = 0$.

Exercise 2: Determine the interval of convergence for the power series $\Sigma((x-3)^n/n)$ from n=1 to infinity.

Solution: To find the interval of convergence, we can use the Ratio Test, which states that a series Σa_n converges if the limit as n goes to infinity of $|a_{n+1}/a_n|$ is less than 1. In this case, we have $a_n = (x-3)^n/n$. Therefore, we find:

$\lim(n->\infty) |(x-3)^{(n+1)}/(n+1) * n/(x-3)^n| = |(x-3) * n/(n+1)|$.

As n goes to infinity, this limit becomes $|x - 3|$, which must be less than 1 for convergence. Therefore, the series converges for $2 < x < 4$.

Problem-Solving Techniques

In the problems above, we applied two important problem-solving techniques for working with power series and Taylor series: constructing the series using the definition, and finding the interval of convergence using the Ratio Test.

When constructing a power series or a Taylor series, the most important steps are finding the nth derivative of the function at the center point c, and then plugging this value into the series formula. This often requires knowledge of derivatives and algebraic manipulation.

When finding the interval of convergence, the Ratio Test is often the most useful tool. However, it's also important to check the behavior of the series at the endpoints of the interval, as the series may converge at one or both endpoints even if the Ratio Test is inconclusive.

Through repeated practice and application of these techniques, you will become more comfortable with the concepts of power series and Taylor series, and their many uses in calculus.

CHAPTER 9

Calculus in Multiple Dimensions

9.1. Introduction to Multivariable Calculus

Multivariable calculus, also known as calculus of several variables or multivariate calculus, extends the principles of single-variable calculus —derivatives and integrals—to higher dimensions. This area of calculus handles functions of two or more variables in three-dimensional space and beyond, widening the scope of problems that can be solved using these mathematical tools. Understanding multivariable calculus opens the door to a plethora of applications in physics, engineering, computer science, economics, and more.

Understanding Multivariable Functions

At the heart of multivariable calculus are multivariable functions. These are functions that take multiple inputs and produce one or more outputs. The most common type of multivariable function we deal with in this area of calculus is a scalar function of several variables, denoted as $f(x, y, z, ...)$. For example, $f(x, y) = x^2 + y^2$ is a function of two variables.

The graphical representation of such functions requires a three-dimensional coordinate system. For instance, in the case of the function $f(x, y) = x^2 + y^2$, we have the x and y inputs represented on two of the axes, while the output $f(x, y)$ is represented on the third axis.

Vector Fields

One common way to visually represent multivariable functions is through the use of vector fields. A vector field in two or three dimensions is a function that assigns to each point (x, y) or (x, y, z) a two-dimensional or three-dimensional vector. They are a fundamental tool in visualizing and understanding multivariable functions and their gradients (we'll get to gradients later on).

Limits and Continuity in Multivariable Calculus

Just as in single-variable calculus, the concepts of limits and continuity play a significant role in multivariable calculus. A limit of a multivariable function at a certain point is the value that the function approaches as its variables approach that point. The limit does not always exist, but if it does, we can use it to define the continuity of the function at that point. A function is said to be continuous at a point if the limit of the function at that point exists and equals the value of the function at that point.

Partial Derivatives

As we delve deeper into multivariable calculus, we encounter partial derivatives—a key concept. Unlike in single-variable calculus, where a function has just one derivative, in multivariable calculus, a function has multiple partial derivatives—one for each variable the function is defined in terms of. The partial derivative measures how the function changes when we vary just one of the input variables while keeping all other input variables constant.

For example, given a function $f(x, y) = x^2 + y^2$, the partial derivative of f with respect to x (denoted $\partial f/\partial x$) measures how f changes as x changes, while y is kept constant. Similarly, $\partial f/\partial y$ measures how f changes as y changes, with x held constant.

9.2. Partial Derivatives

The concept of a derivative, which measures the rate at which a function is changing at a given point, becomes slightly more nuanced when extended from single-variable calculus to multivariable calculus. In this new context, we deal with partial derivatives, which describe how a function changes as one variable is altered while all other variables are held constant.

Definition of Partial Derivatives

Consider a function of two variables, $f(x, y)$. The partial derivative of this function with respect to x, denoted as $\partial f/\partial x$ or f_x, is defined as the limit:

$$\partial f/\partial x = \lim (h\text{->}0) [f(x+h, y) - f(x, y)] / h$$

This definition mirrors the definition of a derivative in single-variable calculus, except that the variable y is treated as a constant during the calculation. Similarly, we define the partial derivative with respect to y while holding x constant, denoted as $\partial f/\partial y$ or f_y.

To calculate these derivatives, we simply treat all other variables as constants and apply the rules of differentiation that we are familiar with from single-variable calculus.

Geometric Interpretation of Partial Derivatives

The partial derivatives $\partial f/\partial x$ and $\partial f/\partial y$ have geometric interpretations that help in visualizing their significance. In the three-dimensional graph of a function $f(x, y)$, the partial derivative $\partial f/\partial x$ at a point (a, b) represents the slope of the tangent line to the curve obtained by intersecting the surface $z = f(x, y)$ with the vertical plane $x = a$. Similarly, $\partial f/\partial y$ represents the slope of the tangent line to the curve obtained by intersecting the surface with the plane $y = b$.

Higher-Order Partial Derivatives

Just as we can take the derivative of a derivative in single-variable calculus, we can also compute higher-order partial derivatives in multivariable calculus. For a function $f(x, y)$, the second partial derivatives are $\partial^2 f/\partial x^2$, $\partial^2 f/\partial y^2$, $\partial^2 f/\partial x\partial y$, and $\partial^2 f/\partial y\partial x$. These are the derivatives of the first-order partial derivatives with respect to x, y, y, and x respectively. It's important to note that, under mild conditions (if the function's mixed partial derivatives are continuous), the order of differentiation does not matter, i.e., $\partial^2 f/\partial x \partial y = \partial^2 f/\partial y \partial x$, a property known as Clairaut's theorem.

The Gradient Vector

The partial derivatives of a function f(x, y) can be combined into a vector called the gradient vector, denoted by ∇f or grad f. The gradient of f is a vector that points in the direction of the greatest rate of increase of f, and its magnitude is the rate of increase in that direction. For a function f(x, y), the gradient is given by $\nabla f = [\partial f/\partial x, \partial f/\partial y]$.

9.3. Multiple Integrals

In multivariable calculus, we generalize the concept of a single integral to higher dimensions. This section introduces multiple integrals, starting with double and triple integrals, and provides methods for computing them.

Double Integrals

A double integral allows us to integrate a function of two variables over a region in the xy-plane. If we have a function f(x, y) and a region R in the plane, the double integral of f over R is denoted as $\iint_R f(x, y) \, dA$, where dA represents a small area element in the plane.

Conceptually, the double integral calculates the volume of the solid that lies below the surface z = f(x, y) and above the region R in the plane. If f(x, y) gives the density at the point (x, y), the double integral gives the total mass of the material distributed over the region R.

To evaluate a double integral, we typically use iterated integrals. We integrate first with respect to one variable, treating the other as constant, and then integrate the resulting function with respect to the other variable.

Triple Integrals

Triple integrals extend the concept of integration to three dimensions. Given a function f(x, y, z) and a region V in space, the triple integral of f over V is denoted as $\iiint$V f(x, y, z) dV, where dV represents a small volume element in space.

The triple integral calculates the hypervolume under the hypersurface described by w = f(x, y, z) and above the region V in space. If f(x, y, z) gives the density at the point (x, y, z), the triple integral provides the total mass of the material distributed over the region V.

Evaluating a triple integral involves performing an iterated integral three times, first with respect to x, then y, and finally z (or in any order that is convenient based on the function and the region).

Fubini's Theorem

Fubini's theorem is a result that ensures the order of integration does not affect the value of a multiple integral, provided the function being integrated is continuous over the region of integration. In essence, it guarantees that we can compute multiple integrals as iterated integrals.

Change of Variables

Sometimes, a multiple integral can be made easier by changing the variables, analogous to substitution in single-variable calculus. The most common such change of variables for double integrals is to polar coordinates (r, θ), while for triple integrals, it's to cylindrical coordinates (r, θ, z) or spherical coordinates (ρ, θ, φ).

9.4. Vector Calculus

Vector calculus is the branch of calculus that deals with vector fields. This section introduces fundamental concepts of vector calculus, including gradient, divergence, curl, line integrals, surface integrals, and the two major theorems of vector calculus - Green's Theorem and Stokes' Theorem.

Gradient

The gradient of a scalar field f(x, y, z) is a vector field whose value at a point is the vector of the maximum rate of change of f at that point. If we denote the gradient operator by ∇ ("nabla"), then the gradient of f, denoted by ∇f or grad f, is (∂f/∂x, ∂f/∂y, ∂f/∂z). Each component of the gradient represents the rate of change of f in the direction of the corresponding coordinate axis.

Divergence

The divergence of a vector field F(x, y, z) = (P, Q, R) is a scalar field that measures the "outward flux" of F per unit volume. The divergence of F, denoted by ∇·F or div F, is ∂P/∂x + ∂Q/∂y + ∂R/∂z. In physical terms, divergence at a point measures the magnitude of the source or sink at that point.

Curl

The curl of a vector field F(x, y, z) = (P, Q, R) is another vector field that measures the "circulation" or "rotation" of F. The curl of F, denoted by ∇×F or curl F, is a vector whose components are partial derivatives of P, Q, and R. In physical terms, the curl at a point is a vector pointing in the direction of the axis of rotation of F at that point, with magnitude equal to the rotational speed.

Line Integrals

A line integral integrates a scalar or vector field along a curve in space. If f(x, y, z) is a scalar field and C is a curve parameterized by r(t), then the line integral of f along C is ∫C f dr. If F(x, y, z) is a vector field, then the line integral of F along C is ∫C F·dr, which computes the work done by the field F along the curve C.

Surface Integrals

A surface integral integrates a scalar or vector field over a surface in space. If f(x, y, z) is a scalar field and S is a surface parameterized by r(u, v), then the surface integral of f over S is ∫S f dS. If F(x, y, z) is a vector field, then the surface integral of F over S is ∫S F·dS, which computes the flux of the field F across the surface S.

Green's Theorem

Green's theorem relates a line integral around a simple closed curve C to a double integral over the plane region D bounded by C. It states that if F = (P, Q) is a vector field on D, then the line integral of F around C equals the double integral of the curl of F over D.

Stokes' Theorem

Stokes' theorem generalizes Green's theorem to a surface integral over a surface S bounded by a simple closed curve C. It states that the surface integral of the curl of a vector field F over S equals the line integral of F around C.

9.5. Exercises and Problem-Solving Techniques

In the final section of Chapter 9, we focus on exercises and problem-solving techniques that can enhance your understanding of multivariable calculus. By solving these exercises, you can apply the theoretical concepts you've learned, such as vector fields, gradients, divergence, curl, line integrals, surface integrals, and the theorems of Green, Gauss, and Stokes.

Visualizing Vector Fields

Understanding how to visualize vector fields is important in multivariable calculus. Drawing simple vector fields by hand, using arrows to represent the direction and magnitude of the vectors at various points in the space can be helpful. When dealing with more complex fields, computational tools, such as Mathematica or Python's Matplotlib, can provide accurate 2D and 3D plots.

Exercise: Sketch the **vector** field F(x, y) = xi - yj and identify areas of the field where the vectors are pointing upwards, downwards, towards the right, and towards the left.

Computing Gradients, Divergence, and Curl

Given a scalar field or vector field, you should practice computing its gradient (if it's a scalar field), or its divergence and curl (if it's a vector field). Remember, the gradient of a scalar field points in the direction of greatest increase of the function, the divergence of a vector field measures how much of the vector field is diverging from a point, and the curl of a vector field measures the rotation of the vector field around a point.

Exercise 1: Given the scalar field $f(x, y, z) = x^2 + y^2 + z^2$, compute its gradient. Then given the vector field $F(x, y, z) = xi + yj + zk$, compute its divergence and curl.

Evaluating Line and Surface Integrals

Practicing line and surface integrals are essential. For line integrals, you would need a parametrization of the curve, and for surface integrals, a parametrization of the surface. Then, you would evaluate the integral as you would a single-variable integral, using the parametrization to transform the integral into a more familiar form.

Exercise 2: Evaluate the line integral of $F = yi - xj$ over the path from $(0,0)$ to $(1,1)$ along the line $y=x$. Then evaluate the surface integral of $F = xi + yj + zk$ over the hemisphere $x^2 + y^2 + z^2 = 1$, $z >= 0$.

Applying the Fundamental Theorems

It's also crucial to understand when and how to apply Green's, Gauss', and Stokes' theorems. Each of these theorems reduces a harder problem to an easier one, provided that certain conditions are met. For example, Green's theorem allows you to compute a line integral around a simple closed curve C by instead computing a double integral over the region D bounded by C.

Exercise 3: Use Green's theorem to evaluate the line integral of $F = (2xy - y^2)i + x^2j$ around the triangle with vertices $(0,0)$, $(1,0)$, and $(0,1)$.

Real-World Applications

Lastly, try to apply these concepts in a real-world context. Physics is rich with problems that require the use of vector calculus, as many physical quantities, like force, velocity, and electric and magnetic fields, are vector fields.

Exercise 4: Consider a fluid flow described by the vector field $F(x, y, z) = -y\,i + x\,j$. Calculate the circulation of the fluid around the circle $x^2 + y^2 = 1$.

Solving these problems will not only consolidate your understanding of the topics covered but also give you insights into how multivariable calculus can be used to solve complex, real-world problems. Remember, mathematics is not a spectator sport. The best way to learn is by doing, so dive into these exercises and work through them step by step.

Solutions

Exercise 1: Given the scalar field $f(x, y, z) = x^2 + y^2 + z^2$, compute its gradient. Then given the vector field $F(x, y, z) = xi + yj + zk$, compute its divergence and curl.

Solution:

The gradient of a scalar field f is a vector field whose components are the partial derivatives of f with respect to each of the coordinates. For the scalar field $f(x, y, z) = x^2 + y^2 + z^2$, the gradient ∇f is computed as follows:

$\nabla f = (\partial f/\partial x)i + (\partial f/\partial y)j + (\partial f/\partial z)k$

$= (2x)i + (2y)j + (2z)k$

The divergence of a vector field F measures the rate at which "density" exits a given region of space. For the vector field $F(x, y, z) = xi + yj + zk$, the divergence $\nabla \cdot F$ is computed as follows:

$\nabla \cdot F = \partial P/\partial x + \partial Q/\partial y + \partial R/\partial z$

$= \partial/\partial x\,(x) + \partial/\partial y\,(y) + \partial/\partial z\,(z)$

$= 1 + 1 + 1 = 3$

The curl of a vector field F, denoted by $\nabla \times F$, is a vector field that measures the rotation of F. For the given vector field, the curl is computed as follows:

$\nabla \times F = (\,(\partial R/\partial y - \partial Q/\partial z)i, (\partial P/\partial z - \partial R/\partial x)j, (\partial Q/\partial x - \partial P/\partial y)k\,)$

$= (\,0i, 0j, 0k\,)$

$= 0$

So, the gradient of the given scalar field is $(2x)i + (2y)j + (2z)k$, the divergence of the given vector field is 3, and the curl of the vector field is 0.

Exercise 2: Evaluate the line integral of $F = yi - xj$ over the path from $(0,0)$ to $(1,1)$ along the line $y = x$. Then evaluate the surface integral of $F = xi + yj + zk$ over the hemisphere $x^2 + y^2 + z^2 = 1$, $z \geq 0$.

Line Integral Solution:

First, we want to evaluate the line integral of the vector field $F = yi - xj$ over the line $y = x$ from $(0,0)$ to $(1,1)$. A line integral sums up the product of a function with the arc length of a curve.

The path we're interested in is the line segment from (0,0) to (1,1) along the line y = x, so we can parametrize this curve as r(t) = ti + tj for 0 <= t <= 1.

Then we substitute these into our vector field F to get F(r(t)) = tti - ttj = ti - tj. The line integral ∫F.dr over our path is thus ∫(from 0 to 1) F(r(t)).r'(t) dt = ∫(from 0 to 1) (ti - tj).(i + j) dt = ∫(from 0 to 1) t dt = [1/2 * t²]_(from 0 to 1) = 1/2.

Surface Integral Solution:

Now, we want to evaluate the surface integral of the vector field F = xi + yj + zk over the hemisphere $x^2 + y^2 + z^2 = 1$ with z >= 0. A surface integral sums up a function over a surface in space.

We parametrize our hemisphere as φ(u, v) = (sin(u)cos(v))i + (sin(u)sin(v))j + cos(u)k for 0 <= u <= pi/2 and 0 <= v <= 2pi.

Then we find φ_u x φ_v = (sin^2(u)cos(v))i + (sin^2(u)sin(v))j + (sin(u)cos(u))k.

The surface integral ∫∫F.dS over our hemisphere is ∫∫(from u=0 to pi/2 and v=0 to 2pi) F(φ(u,v)).(φ_u x φ_v) du dv = ∫∫(from u=0 to pi/2 and v=0 to 2pi) (sin(u)cos(v)i + sin(u)sin(v)j + cos(u)k).(sin^2(u)cos(v)i + sin^2(u)sin(v)j + sin(u)cos(u)k) du dv = pi/2.

Exercise 3: Use Green's theorem to evaluate the line integral of F = (2xy – y^2)i + x^2j around the triangle with vertices (0,0), (1,0), and (0,1).

Solution:

Green's theorem is a fundamental result in vector calculus that relates a line integral around a simple closed curve C to a double integral over the plane region D bounded by C. Specifically, Green's theorem states that:

∫_C F·dr = ∫∫_D (curl F)·dA

In this exercise, we are given the vector field F = (2xy – y^2)i + x^2j. The curl of this vector field is given by:

curl F = ∇×F = (∂/∂x, ∂/∂y, ∂/∂z) × (2xy – y^2, x^2, 0) = (0, 0, 2x - 2x - (-2y)) = 2yi.

According to Green's theorem, the line integral of F around the triangle with vertices (0,0), (1,0), and (0,1) is equal to the double integral of the curl of F over this triangle. This triangle corresponds to the region D in the xy-plane defined by 0 ≤ x ≤ 1, 0 ≤ y ≤ 1 - x.

So, we have:

∫_C F·dr = ∫∫_D (curl F)·dA = ∫ (from 0 to 1) ∫ (from 0 to 1 - x) 2y dy dx = ∫ (from 0 to 1) [y^2](from 0 to 1 - x) dx = ∫ (from 0 to 1) (1 - x)^2 dx = [x/3 - x^2/2 + x^3/3](from 0 to 1) = 1/3 - 1/2 + 1/3 = 1/6.

So, the line integral of F around the given triangle is 1/6.

Exercise 4: Consider a fluid flow described by the vector field F(x, y, z) = -y i + x j. Calculate the circulation of the fluid around the circle x^2 + y^2 = 1.

Solution:

In vector calculus, circulation is a measure of the "rotation" or "angular momentum" of a vector field along a closed curve. It's defined as the line integral of the vector field along the curve. If the curve is parameterized by r(t) = (x(t), y(t), z(t)), then the circulation of the vector field F = (P, Q, R) around the curve is given by:

$\int F \cdot dr = \int (P\, dx + Q\, dy + R\, dz)$

In this exercise, we have the vector field F = -y i + x j and the closed curve is the circle $x^2 + y^2 = 1$. This circle can be parameterized by r(t) = (cos(t), sin(t)), for t in $[0, 2\pi]$.

Substituting the parametric equations into F gives F(r(t)) = -sin(t) i + cos(t) j.

The differential dr = dx i + dy j can be written in terms of dt as dr = (-sin(t) dt) i + (cos(t) dt) j.

So, we get F · dr = (-sin(t) * -sin(t) + cos(t) * cos(t)) dt = $(\sin^2(t) + \cos^2(t))$ dt = dt.

Therefore, the circulation of the fluid around the circle is:

$\int$ (from 0 to 2π) dt = [t]_(from 0 to 2π) = 2π - 0 = 2π.

This means that the circulation of the fluid around the given circle is 2π.

CHAPTER 10

Next Steps

As we come to the end of this comprehensive exploration of calculus, it is important to reflect on what we've learned and look towards the journey that lies ahead. Through our exploration of limits, derivatives, integrals, differential equations, sequences, series, and multivariable calculus, we've learned that calculus is not just a set of rules and procedures. Instead, it is a powerful language that allows us to understand and describe the world around us in a precise and meaningful way.

The core of calculus—studying change and accumulation—permeates all aspects of life and various fields of study, from physics to economics, from engineering to medicine, and from computer science to social sciences. We've seen how the derivative enables us to quantify rates of change and how the integral helps us calculate accumulated quantities. We've explored sequences, series, and the convergence and divergence of infinite sums. We've delved into the world of multivariable calculus, and we've seen how calculus plays an essential role in numerous practical applications.

Yet, as with all journeys, the completion of one phase simply ushers in the next. While this course has equipped you with a solid foundation, there are numerous directions you can take from here. You might choose to delve deeper into pure mathematics, exploring fields such as real analysis or complex analysis, which formalize and generalize many of the concepts introduced in calculus. You might pursue the study of differential equations further, delving into topics such as partial differential equations or dynamical systems.

If you're interested in the application of mathematics, you might study fields such as physics, engineering, or computer science in greater depth, where the tools of calculus are used to model and solve real-world problems. In economics or social sciences, you might apply the principles of calculus to model and analyze the behavior of individuals and societies. In the realm of computer science and data science, you might use calculus as a foundation for machine learning and optimization.

What's important is that the journey does not stop here. The concepts, techniques, and skills you've learned in calculus are transferable to many other domains and will prove invaluable no matter your path. Calculus is not an endpoint, but a launchpad. It's a crucial component of a broader mathematical education, one that nurtures problem-solving skills, logical reasoning, and the ability to handle abstraction and complexity.

Ultimately, the study of calculus is part of the larger journey towards understanding the mathematical structure of the world. It's a journey of understanding how things change and accumulate, and of appreciating the beauty, power, and elegance of mathematical thought. It's a journey that opens up new ways of seeing and understanding the world, and it's a journey that is well worth embarking upon. As you move forward, may the lessons, skills, and insights gained from your study of calculus serve you well, opening doors to new knowledge and opportunities.

Glossary

Chain Rule: In calculus, the chain rule is a formula to compute the derivative of a composite function. That is, if a variable z depends on the variable y, which itself depends on the variable x, then the chain rule states that one can compute the derivative of z with respect to x by computing the product of the derivative of z with respect to y and the derivative of y with respect to x.

Convergence: Convergence refers to the property of a sequence or series to approach a limit. If the sequence or series approaches a specific number, it is said to be convergent. If it doesn't approach any number, it is divergent.

Derivative: The derivative measures how a function changes as its input changes. In other words, it's a measure of the rate of change or slope of a function.

Differential Equation: A differential equation is an equation that involves an unknown function and its derivatives. They are used to model real-world problems where the rate of change of a quantity is known, but the quantity itself is unknown.

Divergence: Divergence is the opposite of convergence. It refers to a sequence or series that does not approach a specific limit.

Implicit Differentiation: Implicit differentiation is a procedure of finding the derivative of an implicitly defined function. By treating one variable as the "independent variable" and differentiating the other variable implicitly.

Integral: The integral is the inverse operation to differentiation. It is used to calculate areas under curves, and in physics, it's used to calculate quantities accumulated over a period of time.

Limit: The limit of a function is a fundamental concept in calculus that describes the behavior of that function as its argument approaches a certain value.

Multiple Integral: A multiple integral involves integration over a region in more than one dimension. For instance, a double integral integrates a function of two variables over a region in the plane.

Multivariable Calculus: Multivariable calculus is the branch of calculus that deals with functions of multiple variables. It extends concepts of one-variable calculus, such as derivatives and integrals, to multiple dimensions.

Partial Derivative: A partial derivative is the derivative of a function of several variables with respect to one of those variables, with the others held constant.

Power Series: A power series is an infinite series of the form $\sum_{(n=0)}^{\infty}$

Product Rule: The product rule in calculus is a formula used to find the derivative of a product of two or more functions.

Quotient Rule: The quotient rule is a formal rule for differentiating problems where one function is divided by another. It follows from the limit definition of derivative and is given by the formula: if the function $f(x)$ is given by $f(x) = g(x)/h(x)$, then its derivative is given by: $f'(x) = [g'(x)h(x) - g(x)h'(x)]/[h(x)]^2$.

Sequence: A sequence is a list of numbers in a specific order. Sequences can either be finite or infinite.

Series: A series is the sum of a sequence of numbers. Series can either converge (have a finite sum), or diverge (have an infinite or undefined sum).

Taylor Series: In mathematics, a Taylor series is a representation of a function as an infinite sum of terms calculated from the values of its derivatives at a single point.

Vector Calculus: Vector calculus is the branch of mathematics that involves differentiation and integration of vector fields, typically in 3 dimensions.

Further Resources

Understanding calculus can be a challenging task, but with the right resources, anyone can master this powerful mathematical discipline. The following list provides a set of resources which can further deepen your understanding of calculus:

Textbooks

Calculus: Early Transcendentals by James Stewart: This textbook is highly recommended for its clarity and comprehensive coverage of calculus topics.

Thomas' Calculus by George B. Thomas, Maurice D. Weir, Joel Hass: Known for its accuracy, clarity, and applications, this book provides a solid foundation in the principles of calculus.

Calculus by Michael Spivak: This is a more rigorous and in-depth exploration of calculus that is suited for those looking for a challenge.

Online Courses

Khan Academy's Calculus Course: Khan Academy is known for its accessible and comprehensive video courses on a variety of subjects, and its calculus course is no exception.

MIT OpenCourseWare: Massachusetts Institute of Technology offers a number of calculus courses free of charge, ranging from single-variable calculus to multivariable calculus.

Coursera's Calculus Course: Many universities and institutions offer calculus courses through Coursera. These courses often include video lectures, quizzes, and peer-reviewed assignments.

Websites and Online Communities:

Paul's Online Math Notes: This site includes notes, tutorials, and problem sets for a range of math topics, including calculus.

Wolfram|Alpha: A computational engine that can solve calculus problems and plot functions. A great tool for checking your work.

r/learnmath subreddit: This is a community on Reddit where users can ask questions about math problems and concepts, including those related to calculus.

Software and Apps

Desmos: This is a free online graphing calculator that is very useful for visualizing calculus problems.

GeoGebra: GeoGebra offers free online math tools for graphing, geometry, 3D, and more.

1.   ^ Real numbers include all numbers that can be written as a decimal or a fraction, including numbers that may require a decimal expansion, such as 1/3, and numbers that cannot be written as fractions, such as √2, or pi. This set includes all positive numbers, negative numbers, and zero.